T0402576

CONSUMER BEHAVIOUR IN HOSPITALITY AND TOURISM

This insightful and expert-led volume presents a holistic view of the latest, cutting-edge global research on trends and changes in consumer behaviour in hospitality and tourism, with focus on the effects of the COVID-19 pandemic and its impacts on purchase patterns within the industries.

The COVID-19 pandemic has affected every aspect of consumer behaviour, including expenses, ways of shopping, eating, lifestyle, use of technology and individual decision-making processes. This volume presents a carefully selected and logically structured collection of chapters, which aim to identify the factors that influence these new purchase patterns and evaluate how managers, retailers and marketers can develop appropriate strategies that respond to these changes in the market going forward. This book explores the effects of multiple socio-economic factors on individual consumption behaviours and features an array of international case studies.

This book is of pivotal interest for students, scholars and researchers interested in consumer behaviour within the tourism and hospitality industries, as well as providing a useful tool for professionals to develop appropriate strategies that meet the evolving needs of consumers in the market.

Saloomeh Tabari is a Lecturer in Marketing and Strategy at Cardiff Business School, Cardiff University. Her research centres on customer experience, in particular intercultural communication and sensitivity in service and marketing. She has a special interest in issues relating to cultural differences, cross-cultural and cultural centrism, and the provision of service and marketing to different customers to enhance their experiences and perceptions by adopting strategies based upon changes in consumer behaviour. Recently she has also developed an interest in how influencers affect consumer behaviour and decision-making. She co-edited the following books: *Global Strategic Management in the Service Industry: A Perspective of the New Era* and *Celebrity, Social Media Influencers and Brand Performance: Exploring New Dynamics and Future Trends in Marketing*. She is on the Editorial Board of a few international journals and is currently acting as Associate Editor of the *Journal of Islamic Marketing*.

Wei Chen is a Senior Lecturer in Strategic Management at Sheffield Hallam University. He has wide experience in cross-culture management in the hospitality and tourism industry. His book *International Hospitality Management* with Professor Alan Clarke has been published in English, French and Portuguese and he has translated books such as *Trade Show and Event Marketing* into different languages. He is the chief overseas editor for *Finance and Economy*, a business magazine, in China. Wei Chen is also a Senior Business Advisor of a renowned sport company in England.

Nazan Colmekcioglu is a Senior Lecturer (Associate Professor) in Marketing and Strategy at Cardiff University and a Visiting Professor at the University of International Business and Economics in Beijing. Her research interests include consumer behaviour in online and offline environments, particularly focusing on factors such as ethical ideology, culture, religion and emotions of customers to understand consumption and anti-consumption attitudes towards products and services in different industries.

CONSUMER BEHAVIOUR IN HOSPITALITY AND TOURISM

Contemporary Perspectives and Challenges

Edited by Saloomeh Tabari, Wei Chen and Nazan Colmekcioglu

Routledge
Taylor & Francis Group

LONDON AND NEW YORK

First published 2025
by Routledge
4 Park Square, Milton Park, Abingdon, Oxon OX14 4RN

and by Routledge
605 Third Avenue, New York, NY 10158

Routledge is an imprint of the Taylor & Francis Group, an informa business

British Library Cataloguing-in-Publication Data
A catalogue record for this book is available from the British Library

ISBN: 978-1-032-63775-4 (hbk)
ISBN: 978-1-032-63776-1 (pbk)
ISBN: 978-1-032-63777-8 (ebk)

DOI: 10.4324/9781032637778

Typeset in Times New Roman
by SPi Technologies India Pvt Ltd (Straive)

This book is dedicated to my parents, Sarah and Mohsen, thank you for believing in me and making this journey happen, you are my inspiration.

To Daniel, the love of my life and my muse. Thank you for always being there for me and cheering me when I thought I could not finish it.

To a friend who became family, Wei, thank you for being you and for your inspirational words. I cherish our collaborations and our friendship and cannot wait for all our ideas to become reality.
Saloomeh Tabari

Thanks to everyone who helped me so much during this very special time in my life. Special thanks to Jinping and kids: your love makes me strong.
Wei Chen

I dedicate this book to all my former students, whose curiosity, enthusiasm, and thirst for knowledge have been a constant source of inspiration. Your eagerness to learn, your insightful questions, and your relentless pursuit of understanding have fueled my passion for teaching and writing. It is an honour to have been a part of your educational journey, and I am grateful for the opportunity to continue sharing knowledge and insights with you all through this book. May you always approach life with the same curiosity and determination that you brought to my classrooms.
Nazan Colmekcioglu

CONTENTS

FIGURES

TABLES

CONTRIBUTORS

Rebecca Biggins is Associate Dean at York Business School, York St John University. Her research interests are around social anthropological marketing. She is fascinated by generational cohort theory and how our experiences, attitudes and views impact our behaviour as consumers.

Wei Chen is a senior lecturer in strategic management at Sheffield Hallam University. He has wide experiences in cross-culture management in the hospitality and tourism industry. His book 'International Hospitality Management' with Professor Alan Clarke has been published in English, French and Portuguese and he has translated books such as 'Trade Show and Event Marketing' into different languages. He is the chief overseas editor for 'Finance and Economy', a business magazine, in China. Wei Chen is also a senior business advisor of a renowned sport company in England.

David D'Acunto is Assistant Professor of Marketing at the Department of Management, University of Verona, Italy. His research interests include digital marketing, eWOM in the service industry, sustainability, circular economy and consumer behaviour in tourism and hospitality.

Saurabh Kumar Dixit is Professor and founding Head of Department of Tourism and Hotel Management North-Eastern Hill University, Shillong Meghalaya, India. His research interests include consumer behaviour, gastronomic tourism, service marketing, experience management and marketing in tourism contexts.

Michael Donald is a Co-Founder of Halo Business Consulting and specialises in operations and customer experience. Halo supports clients around the globe to drive customer loyalty through customer experience, operations, marketing and communications.

İrem Enser is Lecturer at Yaşar University. She conducts academic research on strategic management, corporate social responsibility and hotel management.

Nicole Hankins attended Radford University in Radford, VA, where she recently obtained her Bachelor of Science degree in Recreation, Parks, and Tourism with a concentration in Tourism and Special Events along with an Undergraduate Research Scholar designation. Her research interests include destination management and visitor management.

Elsie Vezemburuka Hindjou is a PhD scholar at the North-Eastern Hill University in the Department of Tourism and Hotel Management, India. Her research interests include gastronomic tourism, destination branding and sustainable tourism development.

Ece İpekoğlu is Assistant Professor at Yaşar University Vocational School Foreign Trade programme. Her academic studies focus on marketing, consumer behaviour, brand management and current marketing approaches.

Adam Jezierski is a PhD student in the field of Management and Quality Studies at Jagiellonian University. His research areas of interest are tourism and hotel management with a focus on reputation management in an online environment. He is also Co-Manager in a family-owned hotel.

Monisha Juneja is currently pursuing a PhD from Amity University, Noida, India, in Travel and Tourism, and is Assistant Lecturer at State Institute of Hotel Management, Indore, Madhya Pradesh, India. Her primary research areas include cultural heritage, world heritage sites, tourist experience and perception.

Ethilde Tulimuwo Kuwa is Lecturer of Tourism Management at the Namibia University of Science and Technology. Her research areas are eco-tourism and community-based tourism, Heritage and cultural tourism, air transportation, and destination marketing.

Micol Mieli is a researcher in consumer behaviour, with a particular focus on technology, information search and spatiotemporal behaviour. Her focus is on exploring the role of different technologies in the contemporary consumer experience and how digital technologies are shaping consumption.

Brendan Paddison is Associate Professor in Business and Management, at York Business School, York St John University. Brendan leads the Visitor Economy and Experience research group, is co-chair of the Tourism Education Futures

Initiative (TEFI) and chairs the York Tourism Advisory Board. He is a member of the editorial boards for the *Journal of Teaching in Travel and Tourism* and the *e-Review of Tourism Research journal* and is also the Book Review Editor for *Enlightening Tourism - A Pathmaking Journal.*

Akshaya Pawar is a PhD student in Hospitality and Tourism Management in the Isenberg School of Management at the University of Massachusetts Amherst, USA. Her research interests focus on hospitality and tourism strategy, and hospitality finance.

Aurimas Pumputis is a doctoral candidate at Lund University. He researches the platform economy in tourism industry and sociomateriality and is interested in the use of digital platforms for building trust between consumers and service providers and managing their accountability relations.

Chen Ren is Lecturer in Marketing and Strategy, at York Business School, York St John University. Her research interests lie within brand management, including luxury branding consumption, luxury experience consumption, brand crisis management, branding in not-for-profit industry and generational studies, particularly cross-disciplinary branding practicality.

Cristian Rizzo is Senior Researcher at the University of Turin. His research activity mainly focuses on consumer psychology and sustainable consumption behaviour and has published several articles in international peer-reviewed journals, such as *Journal of Business Research, Regional Studies, Journal of Cleaner Production* and *Journal of Services Marketing.*

Oscar Robayo-Pinzon is Principal Research Lecturer at the School of Management and Business, Universidad del Rosario, Bogota, Colombia. He is a Fellow of the Association for Consumer Research (ACR) and the Academy of Marketing Science (AMS) and an academic visitor at Cardiff Business School, Cardiff University. His research interests include digital consumer behaviour, digital marketing, mobile marketing, artificial intelligence, social media marketing and behavioural economics.

Claudia Marcela Rodríguez is a professional in Marketing, Diploma in Digital Community Management, Strategic Marketing, Customer Service Management (standard UNE-CEN 16,800 Excellence in customer service) and has experience in customer service management.

Sandra Rojas-Berrio is Director of the Management and Marketing research group and teaches and conducts research in the field of marketing. She is an Associate Researcher for Minciencias and Associate Professor at the Faculty of Economics, Universidad Nacional de Colombia.

Valerio della Sala is an Urban Geographer and Postdoctoral Research Fellow at the University of Bologna. His research covers topics such as urbanism, human geography, environment, landscape and urban culture, including applied projects on the sport, spatial policy and the retrofitting of urban space and cultural-led regeneration programs.

Eleanor Jayne Scanlon is Food and Beverage Manager at The Crown Hotel Harrogate. She oversees all dining aspects, ensuring each guest experience is exceptional. Eleanor's dedication extends beyond her role, as she mentors aspiring professionals and continues her own education.

Andrea Sestino is Adjunct Professor of Competitive Strategy at the LUISS Guido Carli University (Rome), Research Fellow in Business Management at the University of Rome Tre, and Research Assistant in Management and Innovation in Healthcare Services, and Technology Innovation Governance, at the Catholic University Sacro Cuore (Rome). His research interests are new technologies and the impact on managerial/marketing strategies.

Tahir Sufi is Professor and Deputy Director with Amity School of Hospitality, Amity University, Noida, India. He leads the PhD Program (FRC) of the University. His specialisations include human resources management, hospitality management, strategic management and entrepreneurship. He is on the Editorial Board of the *Journal of Business Strategy, Finance and Management*.

Saloomeh Tabari a Lecturer in Marketing and Strategy at Cardiff Business School, Cardiff University. Her research centres on customer experience in particular intercultural communication and sensitivity in service and marketing. She has a special interest in issues relating to 'cultural differences', 'cross-cultural' and 'cultural centrism' and the provision of service and marketing to different customers to enhance their experiences and perceptions by adopting strategies based upon changes in consumer behaviour.

Ewa Wszendybył-Skulska is Associate Professor of Economic Sciences in Management at Jagiellonian University in Krakow, Poland. Her main areas of research interest are human resource management, quality management, innovation and modern management in hotel and tourism enterprises. She is actively engaged in professional consultancy for hotel entities.

Anita Zatori is Associate Professor at the Department of Recreation, Parks and Tourism at Radford University, Virginia, USA. The primary focus of her research, teaching and consultancy work has been experience design within the field of tourism, events, recreation and hospitality for the past decade. Her professional background is in tour operating, tour guiding and event management.

PREFACE

This collection is designed with researchers whose expertise is aligned with consumer behaviour, including research students with a doctorate in these areas who would like to understand the world of hospitality and tourism. The book is structured to discuss not only the changes in consumption behaviour but also broader considerations of the impact of the Pandemic and after that concerning the new trends within the consumer behaviour within the hospitality and tourism industry. All chapters have been prepared by active researchers in the field and provide a holistic view of current literature and present robust discussion.

This edited book includes 14 chapters covering a range of topics within the contemporary perspective and challenges within the hospitality and tourism industry with regard to the changes in consumer behaviour. In Chapter 1, Valerio della Sala looks at Barcelona and the importance of tourism consumption in the post-pandemic era. In Chapter 2, Ewa Wszendybył-Skulska and Adam Jezierski investigate tourists' experiences in times of uncertainty. In Chapter 3, Nicole Hankins, Akshaya Pawar and Anita Zatori explore the impact of overcrowding on the tourist experience, in a comparison study within Las Vegas and the Great Smoky Mountains. In Chapter 4, Aurimas Pumputis and Micol Mieli discuss the trust and trustworthiness of Airbnb. In Chapter 5, Monisha Juneja and Tahir Sufi explore the impact of trip experiences on overall satisfaction among Indian tourists. In Chapter 6, Elsie Vezemburuka Hindjou, Saurabh Kumar Dixit and Ethilde Tulimuwo Kuwa explore the visitors' consumption in a post-COVID era by looking at the street food market. In Chapter 7, Claudia Marcela Rodríguez, Sandra Rojas-Berrio and Oscar Robayo-Pinzon look at the backpacker's motivation behind choosing the destination. In Chapter 8, Andrea Sestino and Cristian Rizzo explore the travel mood by looking at the psychological flow of culture and adventure travel experiences. In Chapter 9, Eleanor Jayne Scanlon and Saloomeh Tabari look at the perceptions and reactions towards tipping from non-tipping cultures from the employees' expectations. In Chapter 10, David D'Acunto explores the different tourist generations and their behaviour towards luxury buying. In Chapter 11, Saloomeh Tabari and Wei Chen introduce a framework for future small businesses by looking at the impact of the health risk epidemic on purchasing behaviour through the theory of planned behaviour. In Chapter 12, Brendan Paddison, Chen Ren and Rebecca Biggins look at luxury consumption and the need for a unique experience. In Chapter 13, Ece İpekoğlu and İrem Enser explore the effect of social media influencers on food and beverage consumption behaviour. The book ends with Chapter 14, where, Michael Donald provides a discussion on navigating values and expanding the authenticity in the post-pandemic and future of the hospitality and tourism industry as an industry expert.

Saloomeh Tabari, Department of Marketing & Strategy,
Business School, Cardiff University, Wales, UK.

Wei Chen, Sheffield Business School,
Sheffield Hallam University, Sheffield, UK.

Nazan Colmekcioglu, Department of Marketing & Strategy,
Business School, Cardiff University, Wales, UK.

ACKNOWLEDGEMENTS

We would like to thank the contributors to the book for their forbearance in the face of a protracted editorial process, special thanks to the reviewers for their positive and encouraging comments, and the support of the Routledge group.

Saloomeh, Wei and Nazan

INTRODUCTION

Consumer behaviour in hospitality and tourism: Contemporary perspectives and challenges

Saloomeh Tabari, Wei Chen and Nazan Colmekcioglu

Introduction

Welcome to the world of hospitality and tourism, where consumer behaviour shapes the very fabric of the industry. In this edited book, we embark on a journey to explore the contemporary perspectives and challenges that define consumer behaviour in hospitality and tourism. From the bustling streets of urban destinations to the serene landscapes of remote getaways, every interaction between consumers and service providers holds profound implications for the industry's evolution.

As we delve into the intricacies of consumer behaviour, we are confronted with a landscape that is constantly evolving, driven by technological advancements, shifting societal norms, and global events. In this dynamic environment, understanding the motivations, preferences and decision-making processes of travellers is paramount for businesses striving to thrive amidst fierce competition and ever-changing consumer expectations.

Drawing upon insights from psychology, marketing and sociology, this edited book offers a comprehensive exploration of the multifaceted dimensions of consumer behaviour in hospitality and tourism. Through a blend of theoretical frameworks, empirical research and practical case studies, we unravel the complexities of consumer decision-making and examine how businesses can adapt to meet the evolving needs of travellers in the 21st century within the 14 chapters presented in this edition.

Through rigorous analysis and real-world examples, this edited book aims to provide readers with a comprehensive understanding of consumer behaviour in hospitality and tourism. By embracing contemporary perspectives and addressing emerging challenges, businesses can position themselves for success

DOI: 10.4324/9781032637778-1

in an ever-evolving landscape, where the consumer reigns supreme. Moreover, the pandemic has rapidly transformed individual consumption behaviours in the last three years, necessitating an exploration of new trends and changes in consumer behaviour before and after the pandemic (Kim et al., 2022). The fields of hospitality and tourism studies are emerging disciplines (Gursoy, Malodia, & Dhir, 2022; Dixit, 2018), where the body of knowledge is evolving with the advancements in society, culture, technology and human behaviour.

Consumer behaviour in hospitality and tourism is a dynamic and ever-evolving field that offers invaluable insights into the needs, preferences and desires of travellers around the world. As consumption patterns continue to evolve, it is crucial to identify the factors that influence new purchase patterns to help managers, retailers and marketers develop appropriate strategies that respond to these changes in the market. According to Urbanikova et al. (2020), today's consumers are exposed to an increasing range of products and information, leading to an increasing diversity of consumer demand, which challenges service producers to supply the right products according to customer preferences. By understanding the underlying motivations, preferences and decision-making processes of travellers, businesses can anticipate market trends, tailor their offerings, and create truly memorable guest experiences. In this book, we hope to provide readers with a comprehensive understanding of consumer behaviour within the hospitality and tourism sectors, as well as practical strategies for navigating the complexities of this dynamic industry. So, join us as we navigate the intricate nuances of consumer behaviour in hospitality and tourism, unravelling the mysteries that shape the future of the industry.

References

Gursoy, D., Malodia, S., & Dhir, A. (2022). The metaverse in the hospitality and tourism industry: An overview of current trends and future research directions. *Journal of Hospitality Marketing & Management, 31*(5), 527–534.

Kim, J., Yang, K., Min, J., & White, B. (2022). Hope, fear, and consumer behavioral change amid COVID-19: Application of protection motivation theory. *International Journal of Consumer Studies, 46*(2), 558–574.

Urbaníková, M., Štubňová, M., Papcunová, V., & Hudáková, J. (2020). Analysis of Innovation Activities of Slovak Small and Medium-Sized Family Businesses. *Administrative Sciences, 10*, 80.

Dixit, S.K. (2018). The Routledge handbook of hospitality studies. *Hospitality & Society, 8*(1), 99–102.

1

TOURISM CONSUMPTION AS A PRIORITY IN THE POST-PANDEMIC PERIOD

The case study of Barcelona

Valerio della Sala

Introduction

Barcelona is an ideal city to use as a case study for researching an urban tourism management governance model involving territorial stakeholders (Romão et al., 2021). To achieve efficient and effective governance in tourism, it is crucial to establish processes and actions capable of establishing a decision-making process based on participation and collaboration. Participation should involve stakeholders, including tourism industry representatives, residents, government bodies and community organisations (Wray, 2013). Stakeholders should be involved in identifying tourism challenges and opportunities related to the destination and developing effective strategies to address these issues (Lalicic & Önder, 2018). The following research aims to produce knowledge and critically analyse the challenges and opportunities in Barcelona's tourism management. Therefore, the study can provide valuable information to policymakers and tourism practitioners by analysing a best practice and its innovative strategies to cope with the limitations introduced by the COVID-19 pandemic.

In tourism, the city of Barcelona is considered a reference point for international tourism, with a significant tourism brand envied by many destinations (Datzira-Masip & Poluzzi, 2014). Barcelona is recognised as one of the world's major urban tourist destinations (Romão et al., 2021). As we will observe within the contribution, Barcelona's tourism growth has been continuous, except for three moments of crisis: (1) the financial crisis of 2008; (2) Bombing of the Rambla de Catalunya in 2017 and (3) the COVID-19 pandemic (2020–2021). The economic impact of tourism on the city economy in 2019 was equivalent to 12% of tourism GDP (Jutglá, 2019). Furthermore, Barcelona, in the pre-pandemic year, was recognised as the second most visited European

DOI: 10.4324/9781032637778-2

city based on accommodation bookings through apps such as Airbnb, Booking, Expedia and TripAdvisor (EUROSTAT, 2021). According to a study by the International Congress and Convention Association (ICCA), Barcelona is the fourth city in the world regarding congress organisation and the first in terms of number of participants. One of the most significant contributing factors is the transport infrastructure. The marina is the second most important port in Europe in terms of cruise calls and the third in terms of embarkations and disembarkations (Vayá et al., 2018). Josep Tarradellas Airport welcomed more than 50 million passengers in 2019, ranking sixth in Europe in passenger traffic (AENA, 2021). However, excess tourism has introduced new problems and social conflicts, such as protests over the lack of affordable housing. The touristisation of neighbourhoods, the gentrification of commercial activities and the overcrowding of public spaces are all effects that affect the quality of life in neighbourhoods, security, privacy and even local identity (Elorrieta et al., 2022; Garay-Tamajon et al., 2022). Therefore, to address this issue, Barcelona produced a new strategy document for sustainable tourism following the guidance provided during the Glasgow Conference 2021. Barcelona was recognised as the first city in the world to regulate short-term rentals by blocking tourist licences from 2015 to 2017 (Wilson et al., 2022). As we will see, the city of Barcelona and the autonomous community of Catalonia have created collaborative and participatory agencies for the identification of objectives and priorities of all local actors. Therefore, the following elements allow us to state how important the choice of this city as a case study is since it is necessary to observe how the Catalan capital managed to cope with a moment of economic crisis through the reorganisation of tourism policies.

From restrictive measures to gradual reopening in Spain

In Spain, Royal Decree 8/2020 on extraordinary urgent measures to reduce the spread of COVID-19 was promulgated on 17 March 2020. The urgent measures were designed to reduce the contagion spread by containing the country's domestic demand and economic activity. The closure of schools, the suspension of public events and the ban on flights and other public transport substantially impacted demand in the tourism and related services, transport, education and culture sectors. Therefore, the decree reinforced the virus containment measures promulgated in Royal Decree 463/2020 of 14 March, declaring a state of alert for managing the health crisis caused by COVID-19. Containment of the disease's progression involved temporary restrictions on freedom of movement and reduced work activity due to quarantine and containment measures. The following circumstances resulted in a supply and demand shock for the Spanish economy, inevitably affecting companies' sales and generating liquidity tensions that could lead to solvency problems and job losses.

In this context, the top priority for the Spanish government has been to protect and support the productive and social fabric to minimise the impact of the virus. As emphasised by the European Commission in its communication of 13 March, the response to this common challenge had to be addressed synergistically through the support of the EU institutions and the introduction of temporary national measures. In particular, the situation generated by the evolution of COVID-19 led to the need to adopt extraordinary containment measures by public health authorities in the current scenario of reinforced containment, coordinated within the framework of the Interterritorial Council of the National Health System (Consejo Interterritorial del Sistema Nacional de Salud).[1] The temporary measures were aimed at reinforcing the National Health System throughout the Iberian territory, guaranteeing the supply of goods and services necessary to protect public health, food supply, electricity, petroleum products and natural gas, and some measures in the transport sector. Finally, the decree determined the suspension of procedural deadlines and administrative activities.

Evolution of pandemic crisis management in Spain

On 14 March 2020, the Spanish government declared a state of alert throughout Spain to deal with the health emergency caused by COVID-19. The state of alert is updated every 15 days and is initially extended until 00:00 on 21 June 2020.

Subsequently, the Plan de deescalate was promulgated on 28 April 2020, reducing the measures introduced in the declaration of the state of alert concerning the number of people in intensive care. During the so-called new normal phase, urgent prevention, containment and coordination measures are taken to allow the pandemic to continue to be addressed and controlled.

The plan remains in place until 25 October 2020, when the government again declares a national state of alert to contain the spread of COVID-19 SARS infections. The initial duration of the new alert state is extended until 00:00 hours on 9 May 2021. From this date, the Spanish government will no longer declare a state of alert, and each autonomous community will independently manage the different measures to contain and spread the virus.

The de-escalation plan

The main objective of the De-escalation Plan is to ensure that, while keeping public health protection as a benchmark, daily life and economic activity are gradually restored, minimising the epidemic risk to the population's health and avoiding the collapse of the National Health System. In other words, maximum health security combined with social and economic well-being recovery.

The parameters, the values of which are necessary to progress through the various stages of the de-escalation plan and for which continuous monitoring is required, are analysed centrally to assess the intensity and speed of the exit phase:

- Public health, based on data assessing the four strategic capacities already identified and the evolution of the epidemiological situation.
- Mobility (both internal and international) is closely linked to a possible increase in the risk of infection.
- Social dimension (impact of the disease, confinement and de-escalation on the most vulnerable social groups, particularly the elderly).
- Economic activity (assessment of the situation by sector, especially those with the most incredible capacity to carry on and those most affected by the crisis).

The phases of the de-escalation plan:

1 Phase 0, or De-escalation preparation phase:
  - This phase refers to the initial situation, characterised by establishing standard relief measures throughout the country. As the contagion curve reduces, specific activities, especially in the private sphere, are resumed, always in compliance with safety instructions based on the responsibility and self-protection of citizens (individual non-contact sports activities and walking, care of family gardens, certain economic activities with capacity control, etc.).
  - To this end, specific measures are planned in the following phase for outings and physical activity carried out individually from 2 May. In addition, some small economic activities are reopened.
  - For example, the opening of premises and establishments by appointment for the individual attention of customers, such as the opening of restaurants with takeaway food service, without consumption on the premises. Another example concerns the resumption of individual training for professional and federated sportsmen and sportswomen and the basic training of professional leagues.
  - During Phase 0 (pre-de-escalation phase), the preparation of all public premises is intensified with signage and protective measures to prepare for the start of the next phase, Phase I, which is Phase II.
  - In addition, during this phase, measures may be taken that affect only certain territories. In particular, islands with no external mobility and practically zero infection rates. As a result, the island of Formentera in the Balearic Islands and the islands of La Gomera, El Hierro and La Graciosa in the Canary Islands anticipate their de-escalation by a few days to 4 May, placing them directly in Phase I.

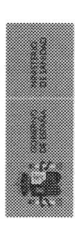

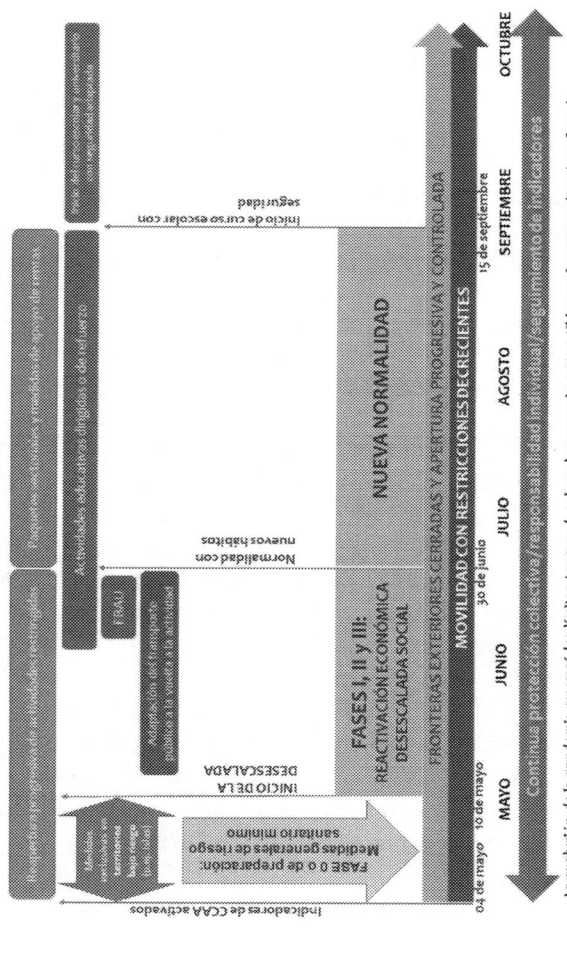

FIGURE 1.1 Descalada Plan, Ministry of Health, 28 April 2020 (Ministerio de Sanidad, 2020).

2 Phase I or Initial:
   - Based on compliance with the scorecard indicators in the different territories, the partial opening of activities is allowed, in particular, economic activities, such as the opening of small businesses; in restaurants, the opening of terraces with occupancy limits of 30%; in hotels and tourist accommodation, opening excluding common areas and with certain restrictions; agricultural activities, professional sports activities with the necessary hygiene measures and places of worship can open with a limit of one-third of their capacity.
3 Phase II or Intermediate:
   - In this phase, activities that have been restricted in phase I can be partially opened. With capacity limitations, the interior space is opened in the restaurant industry with an occupancy of one-third of the capacity and separation guarantees and only for table service. The school year starts in September; however, in Phase II, exceptions are made to reopen educational centres for three purposes: reinforcement activities, ensuring that children under the age of six can attend the centres if both parents have to work in person and the conduct of the baccalaureate assessment for university admission. The resumption of hunting and sport fishing, the reopening of cinemas, theatres, auditoriums and similar venues with pre-assigned seating, and visits to monuments and other cultural facilities, such as exhibition halls and lecture halls, at one-third of their usual capacity, are also planned. Cultural events and performances for less than 50 persons may be held in enclosed venues with one-third of their usual seating capacity, and if they take place outdoors, they may be held when they bring together less than 400 seated persons. Places of worship must limit their capacity to 50% in this second phase.
4 Stage III or Advanced:
   - In this, the last, general mobility is made more flexible, although the recommendation to wear masks outside the home and on public transport is maintained. In the commercial area, capacity is limited to 50% and a minimum distance of 2 metres is established. In the restaurant business, capacity and occupancy restrictions are relaxed, although strict separation conditions between the public are maintained.

As observed in the following phases, limited mobility significantly impacted economic activity. Important lessons were learnt from the following stages, such as the individual responsibility to comply with rules and recommendations or the co-responsibility developed by companies and workers to ensure maximum protection in the workplace. After the following exceptional period and the great effort made, it was necessary to gradually reactivate economic activity in tourism, culture, accommodation and catering sectors. Internationally, the return to normality significantly impacted demand recovery for certain goods and services, especially tourism-related ones. Therefore, the gradual opening of borders required much cooperation and multilateralism so that international tourism could return to normal for the national economy. All the following steps and actions laid the foundations for a secure approach to open borders according to the evolution of the pandemic and prioritising coherence at the European level.

New normality

Social and economic restrictions have ended, but epidemiological surveillance strengthened health system capacity and maintained public self-protection.

The specific dates and actual evolution depend on the behaviour and control of the pandemic and the ability to pass through the different phases under the conditions set out in the plan. The time between each phase has a minimum duration of two weeks, the average virus incubation period.

Looking at the different phases of the de-escalation plan proposed by Spain, from the first phase, citizens in the autonomous communities could carry out certain activities to improve their psycho-physical well-being. After only 50 days, citizens were, in fact, free to engage in motor and sports activities, visit a museum, visit a place of worship, access the library loan service and purchase certain goods and services according to the limitations of the individual autonomous communities. Spain's measures from the outset were designed to improve citizens' well-being, reducing the anxiety and stress that arose during the initial period of the pandemic.

The measures contained in the de-escalation plan promote the practice of motor and sports activities starting on 2 May through a time division and segmented according to the age of the population.[2]

- For non-professional walks and sports (over 14 years of age): 06:00 to 10:00 and 20:00 to 23:00.
- For dependent persons who need to go out accompanied or are over 70 years of age: 10:00 to 12:00 and 19:00 to 20:00.
- For children under 14 (accompanied): 12:00 to 19:00.

Exceptionally, persons who cannot exercise during the established time slots due to medical prescriptions or work–life balance problems may do so at other times with appropriate accreditation. Furthermore, activities may only be carried out in the company if the persons are cohabiting; otherwise, they must be carried out alone.

As noted above, the permitted walking distance must be a maximum of 1 kilometre from the place of residence, and in the case of individual sports, it is forbidden to leave the municipality of residence. Violating the rules imposed by the alert state is subject to economic sanctions.[3]

It is interesting to note how Spain has always spoken of respecting safety distances to limit the spread of the virus, never mentioning social distancing in the various decrees promulgated.

From the very beginning, the Spanish government sought to implement measures that would reduce marginalisation and social exclusion so that the chances of developing psychological problems in the weakest individuals would be reduced.

Furthermore, during this initial period, the use of protective equipment in open spaces is not mandatory if safety distances are observed.

Since adopting the new de-escalation plan, the autonomous communities are responsible for updating the restrictive measures and opening up the planned activities within each phase.

The Catalunya framework

The Catalan Tourism Agency (ACT) is the body of the Government of Catalonia responsible for the implementation of tourism promotion policies within the territory.[4] The ACT, since 2010, replaced the former consortium Turisme de Catalunya, which until then had been responsible for the promotion and implementation of the Autonomous Community's tourism policies.

The Catalan Tourism Agency was created as a public law body in 2007 by the Generalitat to promote Catalonia as a tourist destination of reference.

Following Article 171 of the Statute of Autonomy of Catalonia, the ACT has exclusive competencies in tourism quality and social and economic viability. The transformation of the consortium into an agency guarantees autonomy in promoting Catalonia's attractions abroad and represents the first breakthrough in Catalonia's promotion strategy. One of the significant innovations and challenges that the ACT wants to undertake is the co-management and co-participation of the private sector in promoting and selling the Catalan tourist offer worldwide.

Through the Strategic Plan for Tourism in Catalonia 2018–2022 (PETC), the Agency established the basis for managing tourism activity in Catalonia in view of the 2030 sustainable development goals. According to the World Tourism Organisation, 1,800 million annual tourists are expected in 2030. Therefore, starting from this figure, the Strategic Plan introduces several scenarios of what the international tourist volume Catalonia would like to receive in the future could be. One of the scenarios sets the figure at 35.6 million international tourists in 2030, and within this framework, the PETC makes a sustained progression that puts the figure at around 21 million tourists in 2022. However, the COVID-19-induced pandemic meant that this figure remained only a forecast.

The Strategic Plan aims to position Catalonia as one of the best destinations in the Mediterranean and to achieve the following objectives in 2025:

1. Increase daily expenditure per tourist.
2. Reach 37% of tourists visiting us in March, April, May and October.
3. Increase the number of tourists in inland destinations by up to 10%.
4. Manage growth intelligently to reach 21 million international tourists.

The PETC establishes seven main lines of action developed through 29 initiatives to achieve the following objectives. Furthermore, the PETC was elaborated in collaboration with representatives of the Catalan tourism sector who participated in the annual drafting of the plan.

The seven strategic lines of action of the plan:[5]

1. Offering an exceptional tourist experience.
2. Win and retain the right customers.
3. Attract the necessary investments.
4. Implement smart tourism best practices.
5. Improving the conditions for competitiveness.
6. Improving the tourism management of the territory and its development.
7. Re-think organisation, management and governance in depth.

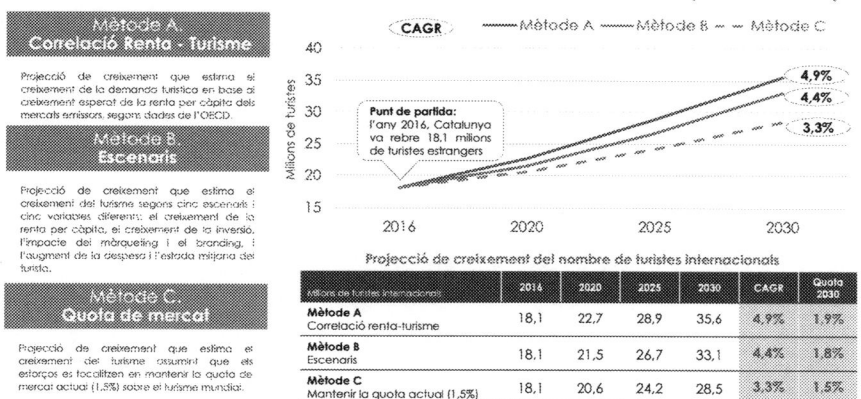

FIGURE 1.2 Comparison of international tourist number growth forecasts for the three methods (in millions of tourists) (THR, based on data from Tourism Economics, 2022).

c. Addicionalment, l'estratègia persegueix 6 objectius quantitatius
■■■

Objectius quantitatius de l'estratègia

Objectiu	Unitat	Estat 2016	Objectiu a assolir 2022	Objectiu a assolir 2025	Variació 2022 vs 2016
1. Nombre de turistes internacionals[1]	Milions de turistes	18	21	24	+17%
2. Despesa mitjana per càpita per dia	€ per persona/dia	162	188	210	+16%
3. Índex d'estacionalitat[2]	% sobre la demanda anual	34	37	40	+9%
4. Índex de desconcentració territorial[3]	% sobre la demanda anual	9	10	12	+11%
5. Índex de competitivitat[4]	Puntuació sobre mil	-	700/1.000	850/1.000	-
6. Índex de turisme intel·ligent[5]	Puntuació sobre mil	-	700/1.000	850/1.000	-

Notes:
- Nota[1]: Assolint la mateixa quota de mercat que l'actual (1,5%) sobre el turisme internacional mundial. (Per a més informació, veure Projeccions de creixement)
- Nota[2]: percentatge de turistes en pre i post temporada, considerant que aquest corresponen als mesos de març, abril, maig i octubre
- Nota[3]: % de turistes en establiments hotelers fora de la costa. La costa inclou Barcelona, Costa Brava i Costa Daurada
- Nota[4]: l'índex de competitivitat s'haurà de crear
- Nota[5]: l'índex de turisme intel·ligent s'haurà de crear

FIGURE 1.3 Quantitative objectives of the strategy (Source ACT, 2017a).

Finally, the ACT strategy aims to pursue six quantitative objectives:

1. Number of international tourists.
2. Median expenditure per capita per day.
3. Seasonality index.
4. Index of territorial deconcentration.
5. Competitiveness index.
6. Index of Intelligent Tourism.

Therefore, from the quantitative objectives, Catalunya for 2025 expects to increase the average daily expenditure of tourists by 16% and a greater devolution of tourist activities. These projections allow us to introduce the specific case of the Catalan community in the following paragraph.

Barcelona framework

The Barcelona Tourism Observatory: City and Region (OTB)[6] is the working platform for statistical information on tourism, knowledge and market information in Barcelona and the rest of the Barcelona province. It comprises

Barcelona City Council, Barcelona Regional Council, Cambra de Comerç de Barcelona and the consortium Turisme de Barcelona. The OTB was publicly presented in June 2017 and is the conclusion completed by the Barcelona City Council, Barcelona Regional Council and Turisme de Barcelona due to the desire to promote knowledge of the tourism market within each institution. Subsequently, incorporating the Cambra de Comerç de Barcelona into the project in 2019 provided a broader vision, reinforcing the transversal and popular nature of the project.

In recent years, the OTB has developed a methodology that serves as a support tool for transferring tourism knowledge to the territory. Furthermore, the creation of the OTB optimises the investment of economic and human resources by integrating the individual projects of each participating entity into broader, shared studies. Thus, the OTB offers an integral vision of Destination Barcelona and the knowledge needed to manage and promote tourism.

OTB's mission is to achieve the following objectives:

1. To generate knowledge by providing analysis, description and evaluation of tourism activity, monitoring it intelligently and transversally.
2. To establish standard methodological criteria for quantifying and characterising tourism activity.
3. Facilitating and accelerating the decision-making process by actively transferring knowledge to stakeholders related to the activity to increase their competitiveness and citizens' quality of life, from academia to public administration, private companies and citizens themselves.
4. Contributing to the sustainability of tourism activity in all its dimensions.
5. To promote the city and the region as an international reference in the observation of sustainable tourism in urban areas and to share knowledge.
6. Actively contribute to the future Tourist Information System of Catalonia.

In 2022, the city of Barcelona again confirmed itself as one of the world's most important tourism and business travel destinations. In 2022, the city recorded 12,463,892 million tourists and 35,344,356 million overnight stays in hotels, hostels, tourist flats and accommodations used for tourism (HUT). Over the past 30 years, demand for hotel accommodation has increased fivefold from 1.7 million tourists in 1990 to 9.1 million in 2018. While overnight stays increased from 3.8 million in 1990 to 19.3 million in 2018.

Therefore, demand has not stopped growing every year. Demand only stopped growing in front of three exceptional moments:

- The financial crisis of 2008.
- Bombing of the Rambla de Catalunya in 2017.
- COVID-19 pandemic in 2020–2021.

Interestingly, the demand for tourist accommodation, through data provided by Barcelona City Open data, registers an increase of 40.8% growth (OTB, 2022a).

Furthermore, according to data from the tourism observatory, one in three tourists chooses tourist accommodation.

Whereas, looking at rural tourism data, there has been a growth of between 25% and 40% since the pandemic. However, it is interesting to note that stays in hotels average 2.1 nights, while those in tourist accommodation average up to 3.75. Finally, hostels recorded an annual occupancy rate of 70.1% in 2022, the highest figure since 1990.

While about the tourism profile of Barcelona, OTB indicates 29.5% domestic tourism against 70.5% international tourism. International tourism is still dominated by a large percentage of French (9.4%), followed by the United States (8.3%), the United Kingdom (6.5%), Germany (5.2%) and Italy (5.0%). When considering tourists' motivation, the data show that 67.7% visit Barcelona for holiday or leisure. The remaining percentage is broken down as

FIGURE 1.4 Data on the demand for tourist accommodation within Barcelona (Observatori del Turisme a Barcelona, 2022).

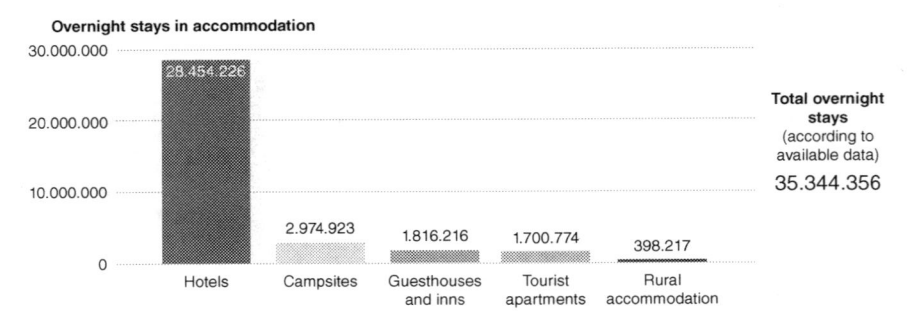

FIGURE 1.5 Data on the number of nights spent in accommodation within Barcelona (Observatori del Turisme a Barcelona, 2022).

18.7% for professional reasons and 13.6% for other reasons. Finally, in 2022, 8 out of 10 tourists reached the city by air transport (OTB, 2022b).

In 2022, the city of Barcelona managed to increase to pre-COVID levels. Concerning the average expenditure per tourist, the data provided by OTB allow us to observe the following average breakdown per day:

- Average transport cost: 361.30 €.
- Average daily expenses: 84.40 €.
- Average cost of accommodation: 68.50 €.

Finally, when looking at the proportion of tourism concerning economic activity, Barcelona has 15.2% of enterprises operating in the tourism sector. At the same time, the employment of tourism human resources represents 12.5% of the total number of employees in Barcelona.

Therefore, the city of Barcelona, although affected by the restrictions of the pandemic, managed to return to a level of employment in 2022, similar to the data observed in the pre-COVID period. As can be seen from the graph (Figure 1.9), the city of Barcelona exceeded 12 million annual tourists and 35 million annual overnight stays in 2022.

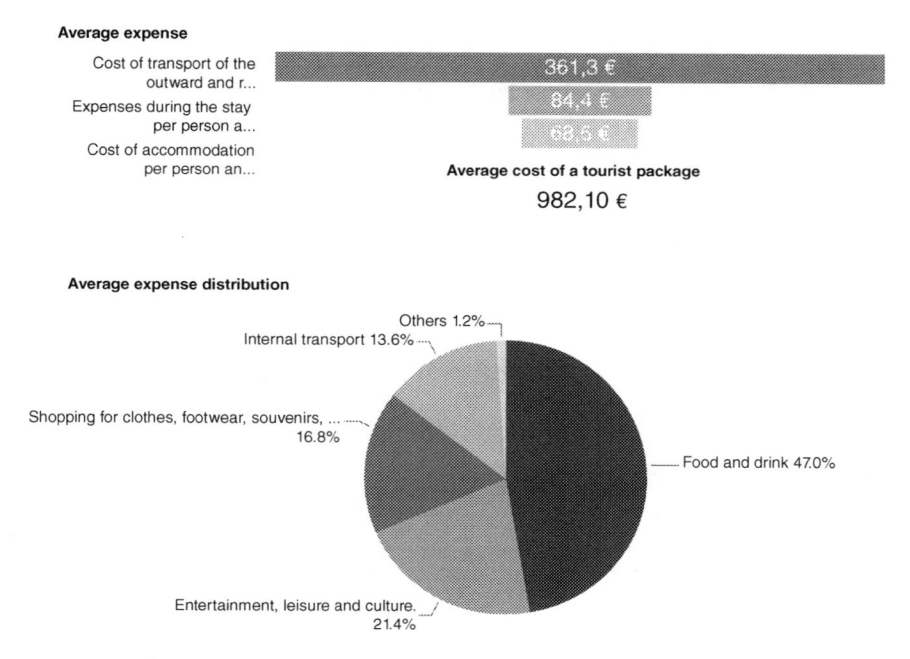

FIGURE 1.6   Average tourist expenditure within Barcelona (source, Observatori del Turisme a Barcelona, 2022).

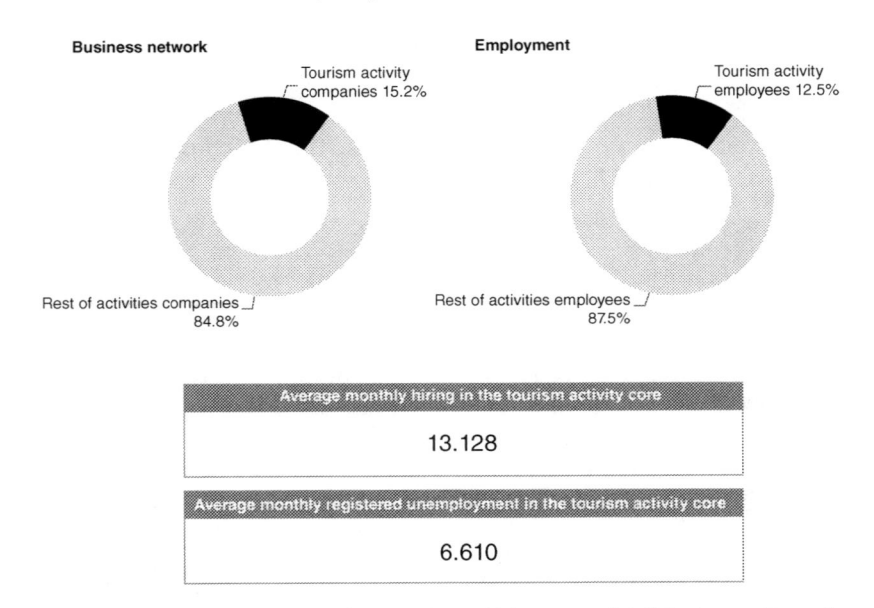

FIGURE 1.7 Labour market within Barcelona (Observatori del Turisme a Barcelona, 2022).

The sustainable tourism strategic plan for the city of Barcelona

Barcelona's strategy for 2025 is to lay the foundations for building a sustainable tourism model based on improving the attractiveness and economic profitability of the tourism sector. Economic profitability will have to consider the natural, cultural, social and political values of the tourist destination to distribute the benefits somewhat over the territory.

The strategy includes a work plan with 16 objectives and 48 actions to be carried out in 2023–2025 that emphasise the Consortium Turisme de Barcelona's commitment to promoting sustainable tourism, understood as a complex system that brings visitors, residents and the region together through integration and collaboration.

The document focuses on three Sustainable Development Goals (SDGs)

- Environmental sustainability.
- Economic sustainability.
- Sociocultural sustainability.

The objectives in the Barcelona City Council's Municipal Tourism Plan include programmes and actions to promote tourism in the city, considering the different effects tourism can have on the local population. Therefore, Barcelona City

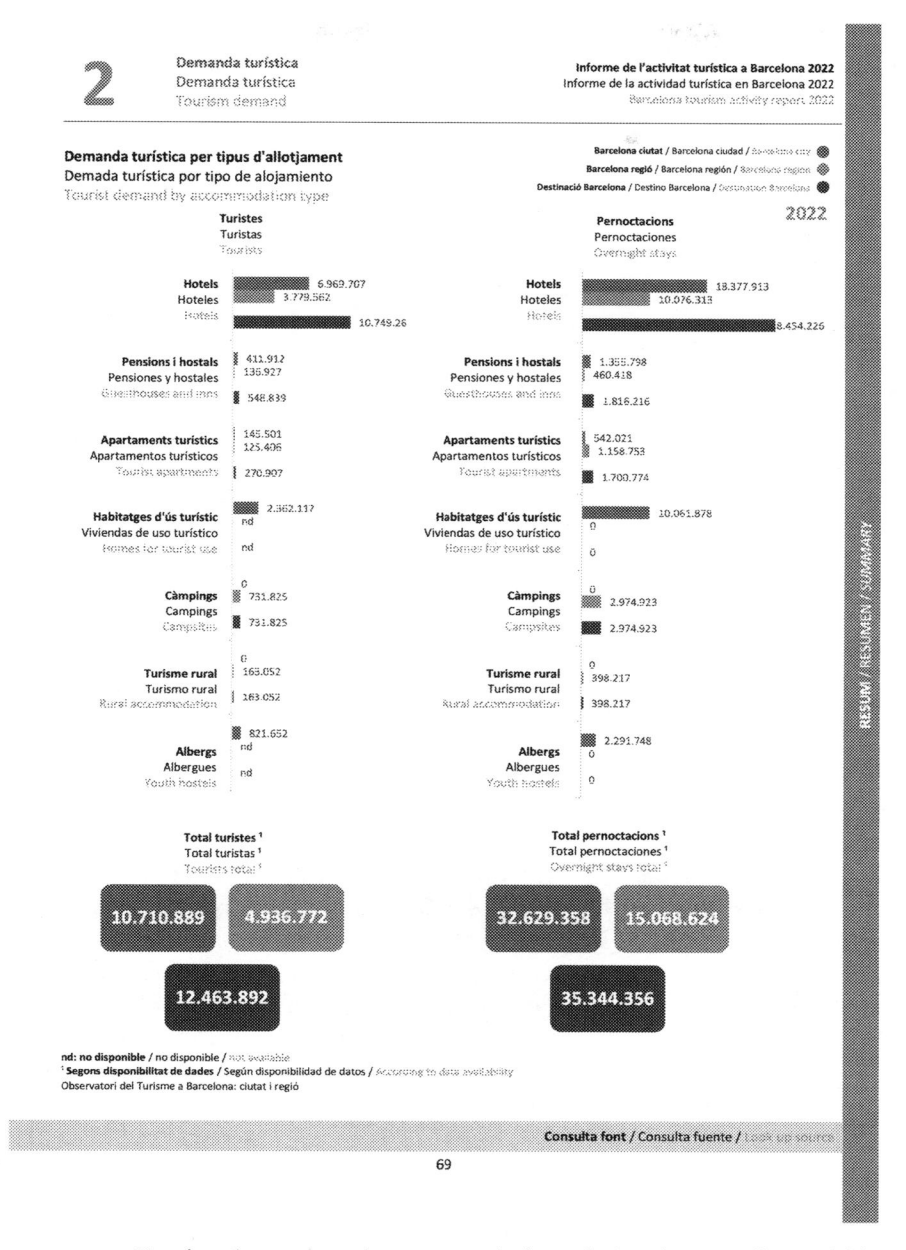

FIGURE 1.8 Tourist demand and accommodation choice by travellers within Barcelona (Observatori del Turisme a Barcelona, 2022).

Council has issued a series of governance plans, agreements and measures that have been taken into account in the implementation of the actions, such as:

- Climate Plan 2018–2030.
- Barcelona Tourism Mobility Strategy.
- Creation of new imagery and content to improve tourist mobility and sustainable tourism.
- Sustainable tourism.
- Tourism marketing strategy for destination Barcelona (EMTDB).
- Barcelona International District Programme.
- Governance measures to promote the Blue Economy in Barcelona.
- Corporate Social Responsibility (CSR).
- Measures to promote gender equality.
- Measures against discrimination.

To make this plan possible, Barcelona Tursime has implemented the Equal Opportunities Action Plan 2021–2025, which sets out guidelines to reduce any incidents of discrimination and combat inequalities in order to guarantee the principle of equal opportunities in all consortium processes. Furthermore, Barcelona in 2011 was the first city in the world to receive the Biosphere World Class Destination certification from the Responsible Tourism Institute, according to the Global Sustainable Tourism Council criteria. The certification, which is renewed annually, recognises Barcelona as a sustainable and responsible tourism destination. In addition, since 2018, companies and entities involved in Barcelona's tourism system that are committed to the responsible management of the environment, culture, working conditions, gender equality and social and economic viability can apply for the sustainability stamp. The process enables the awarding the Biosphere certification for sustainable tourism by the Institute of Responsible Tourism. The certification is entirely subsidised by the occupancy tax levied on hotels, tourist accommodation and all the various forms of accommodation in the area. The following involvement of the city of Barcelona in the project, together with the synergetic collaboration of the public and private sectors, made the city the first destination in the world to receive Biosphere Platinum certification in 2022.

Measures, actions and monitoring of the plan

In 2021, the Consortium Turisme de Barcelona signed the Glasgow Declaration as the supporting organisation for climate action in tourism at the UN COP26 summit.

This meant that the Consortium Turisme de Barcelona is committed to taking actions to support, incentivise and train its members and stakeholders to promote the reduction of the sector's carbon footprint.

The Consortium Turisme de Barcelona actively works in the following areas in order to align with the climate action defined in the declaration:

- *Measure*: Measure and report all travel and tourism-related emissions. Report annually on the measures taken to reduce and offset our greenhouse gas (GHG) emissions.
  - The report on the carbon footprint of tourism in the city of Barcelona was prepared by Inèdit as part of the Strategic Plan for Tourism.
- *Finance*: Stakeholders will have access to the necessary tools and resources to improve their skills and fill gaps in their knowledge.
  - Develop and manage the Barcelona Biosphere Commitment to Sustainable Tourism (BBCST) and Biosphere Certification – action subsidised by the occupancy tax levied on hotels and other accommodation facilities.
  - Carry out campaigns at Turisme de Barcelona to align promotional actions with sustainable tourism and provide free training by experts in the field.
- *Collaborate*: Ensure that tourism contributes to environmental, economic and social balance and considers the resident population.
  - Institutions (Turisme de Barcelona, Chamber of Commerce, City Council and Province) will collaborate with the region's public and private entities and businesses to manage the Barcelona Biosphere Commitment for Sustainable Tourism.
  - Collaborate with stakeholders in sustainable tourism worldwide and initiate a comparative analysis between different cities.
  - Recognise good practices in the sector and promote networking between entities and enterprises.
  - Activate and strengthen the Consortium's Sustainability Table.
- *Decarbonise*: To set and deliver science-based targets to accelerate tourism's decarbonisation.
  - Raise awareness among tourists and visitors and improve communication on the sector's environmental impact to help people choose sustainable and low-impact travel options and tourism activities.
  - Promote nearby markets and longer stays.
  - Create and promote low-impact hotel products in the destination.
  - Develop guidelines for a master plan for the decarbonisation of the tourism sector to be achieved by 2050.
- *Regenerate*: Ensure that tourism contributes to environmental, economic and social balance and supports the resident population. Tourism can help restore the natural, cultural, heritage and identifying elements of the region; generate direct interactions between the host community and the visitor; and facilitate learning, transformation and personal reflection processes.
  - Raise visitors' awareness of the city's tourism footprint.
  - Promoting local food.
  - Promote local culture.

STAKEHOLDERS	
BarcelonaTurisme	
INTERNAL	**EXTERNAL**
• **CONSORTIUM MEMBERS** ✓ Barcelona City Council ✓ The Official Chamber of Commerce. Industry and Shipping of Barcelona ✓ Barcelona Promotion Foundation • **GOVERNING BODIES** ✓ General Council ✓ Executive Committee ✓ Tourism and City Municipal Council • **MANAGEMENT** • **STAFF** ✓ Internal staff ✓ External staff	• **COLLABORATING BODIES** ✓ Institutions ✓ Associations and entities ✓ Guilds and professional associations • **PROGRAMME MEMBERS** • **CLIENTS (BUSINESSES)** • **VISITORS** • **CITIZENS** • **MEDIA** • **SUPPLIERS (BUSINESSES)** • **LOCAL COMMUNITY** ✓ Businesses from the region ✓ Social, environmental, cultural, sporting and economic entities • **SECTOR** ✓ Institutions ✓ Associations and entities ✓ Guilds and professional associations • **GOVERNMENT AGENCIES**

FIGURE 1.9 Consortium's 2017 corporate action plan for stakeholders (Barcelona Turisme, 2023).

The consortium's Sustainability Plan seeks to promote actions and development projects that involve, directly or indirectly, different stakeholders in order to strengthen sustainable policies led by the city's government agencies.

Turisme de Barcelona is part of a complex system that brings together different territorial stakeholders, allowing it to maintain an ongoing dialogue to set objectives and measures.

As a supporter organisation of the Glasgow Declaration, Turisme de Barcelona shares the same goals and commitments as all other signatories and is developing a Climate Action Plan focused on sustainable tourism objectives;

it is also committed to taking actions to promote, support, incentivise and train its members and stakeholders to align with the declaration and its five pathways (Measure, Finance, Collaborate, Decarbonise and Regenerate).

The follow-up and evaluation of the Sustainability Action Plan is crucial to ensure the implementation of the proposed actions and the achievement of the set challenges.

In addition, Barcelona Turisme created the sustainability and management divisions to monitor the specific follow-up actions of the Tourism Sustainability Plan.

Therefore, to ensure the effective implementation of the Consortium Turisme de Barcelona's Sustainability Plan, we propose the follow-up and evaluation actions listed below:

- Quarterly follow-up meetings to assess the degree of implementation of the actions.
- Annual follow-up of actions through follow-up files.
- Annual evaluation report.

Annually, the management team will prepare a follow-up report to assess the degree of implementation of actions and present opportunities for improvement to continue to progress in the sustainable management of the consortium.

Annual follow-up reports may include the following information on the evaluation of results and indicators:

- Level of implementation: total number of actions per division.
- Level of development of actions: to determine the degree of achievement of actions according to the established follow-up indicators.
- Actions according to the follow-up indicators defined in the plan for each action.
- Identify opportunities for improvement to be applied to the plan (new actions, scheduling, new indicators, etc.).

The annual evaluation report will be submitted to the General Division, which will then have to evaluate the proposal of new improvement opportunities to be included in the plan to continue to progress in the sustainable management of the Turisme de Barcelona Consortium.

Conclusion: The new challenges for post-pandemic tourism. A sustainable vision

The city of Barcelona's policies and collaborative actions have been proportioned with the help of the UN's 2030 Sustainable Development Goals. Furthermore, Barcelona has included a unique climate action plan to promote and support sustainable tourism initiatives according to the Glasgow

Declaration signed in 2021. From this point of view, tourism is considered a functional activity for promoting new values, environmental, social and economic and introducing new qualitative indicators for measuring the quality of life within neighbourhoods.

The Plan for the Development of Sustainable Tourism 2023–205 of the City of Barcelona guarantees broad participation and collaboration of the proposed actions. In this way, the City Council, through Barcelona Turisme and The Catalan Tourism Agency, seeks to implement tools to reduce the effects of touristisation and gentrification in neighbourhoods at risk of exclusion and marginalisation.

Considering new models of the contemporary city through promoting new activities and developing new models of sustainable tourism contributes to creating new perspectives for the Catalan capital. In Barcelona, the evolution of segregation and gentrification in both central and peripheral neighbourhoods has forced the City Council to react so that the city does not become a mere space to be designed at will.

The interventions implemented in the city's public space through promoting a healthy lifestyle have allowed people to connect with the community and coexist with its diversity. Through the valorisation of cultural, historical, artistic and architectural heritage, the municipality recognised the role of local authorities as a responsible mediator in transmitting values related to environmental, social, and economic sustainability.

Tourism must be valorised and interpreted as an activity capable of generating social capital, taking into account the enterprises, organisations and partners that will be the responsible mediators in the transmission of sustainable values.

Thanks to the cooperation of local stakeholders, the municipality will be able to connect tourists with the community. Promoting well-being in the neighbourhoods and the citizens' sense of belonging will reduce gentrification and touristisation. In this way, the territorial authorities will support the City Council in promoting specific objectives and actions for promoting sustainable tourism, understood as a complex system that brings together values and active actors.

Barcelona Turisme expects the sustainable development model to promote tourism and the city as an element capable of providing equity and social justice, favouring the construction of the model of society we desire. A holistic city model that can reduce the possibilities of social exclusion and sponsors a quality tourist offer for visitors without any exception or inequality.

Collaboration is only a first step that begins with previous planning in which the objectives present in Barcelona's tourism planning strategy will be established.

The chapter discovers how public policies respond to conflicts arising from tourism and what measures the city of Barcelona has introduced to counter the crisis induced by the COVID-19 pandemic.

The following chapter assumes the relevance of tourism research for developing sustainable tourism policies. In this context, the city of Barcelona's strategic documents are analysed to achieve a social impact that can be highlighted and applied in other organisations. This contribution emphasises the importance of using Barcelona as a case study for the progress and improvement of our cities affected by touristisation or gentrification. This study sought to contribute to knowledge in this field, currently so essential but at the same time so little explored. It proposes an urban co-governance model that involves all stakeholders and can support a holistic model of sustainable tourism development. However, tourism has some specific characteristics as an evolving activity that depends on different behavioural-decision-making factors on the part of visitors. While waiting for the results of the sustainable tourism development plan for the Catalan capital, the following case study allows us to observe the uniqueness and particularity of Barcelona to cope with the problems induced by mass tourism over the last 20 years.

The Barcelona municipality's experiences can be considered good practices but they should be considered a flawed model.

Notes

1 Royal Decree-Law 8/2020 of 17 March on urgent extraordinary measures to cope with the economic and social impact of the COVID-19: https://www.boe.es/eli/es/rdl/2020/03/17/8/con.
2 The following time slots do not apply to municipalities with less than 5,000 inhabitants. These localities represent 12% of the Spanish population and can leave between 6 a.m. and 11 p.m.
3 On 14 July 2021, the Constitutional Court declared the measures and sanctions adopted during the state of alert unconstitutional, creating a legal precedent that will force the various economic communities to compensate their citizens.
4 Catalan Tourism Agency (ACT): https://act.gencat.cat/conoce-la-agencia-catalana-de-turismo/?lang=es.
5 Pla estratègic de Turisme de Catalunya: https://act.gencat.cat/wp-content/uploads/2019/01/Resum-PETC.pdf.
6 Observatori del Turisme a Barcelona: https://www.observatoriturisme.barcelona/en/otb-0.

References

ACT. (2017a). Pla estratègic de marketing turístico de Catalunya 2018–2022. Retrieved from: https://act.gencat.cat/que-somos/plan-de-marketing-turistico-de-catalunya-2018-2022/?lang=es
ACT. (2017b). Pla estratègic de turisme de Catalunya 2018–2022. Retrieved from: https://act.gencat.cat/qui-som/pla-estrategic-de-turisme-2018-2022/
AENA. (2021). Barcelona en cifras. Retrieved from: www.aena.es/es/aerolineas/aeropuertos-y-destinos/nuestros-aeropuertos/josep-tarradellas-barcelona-el-prat.html (assessed 13 November 2021).
Ajuntament de Barcelona. (2017). Barcelona Tourism for 2020. A collective strategy for sustainable tourism. Plan Estrategic. Retrieved from: https://ajuntament.barcelona.cat/turisme/sites/default/files/barcelona_tourism_for_2020.pdf
Ajuntament de Barcelona. (2019). Carbon Footprint of Tourism sector in the city of Barcelona. Retrieved from: http://hdl.handle.net/11703/115214

Barcelona Turisme. (2023). Sustainable tourism strategy. Retrieved from: https://barcelonaturisme.com/uploads/web/bst/EstrategiaTurismeSostenibleBarcelonaTurisme23-25_ENG.pdf

European Commission. (2020). Commission Recommendation (EU) 2020/403 of 13 March 2020 on conformity assessment and market surveillance procedures in the context of the COVID-19 threat (GU L 79 del 16.03.2020). Retrieved from CELEX: https://eur-lex.europa.eu/legal-content/EN/TXT/?uri=CELEX:32020H0403

Datzira-Masip, J., & Poluzzi, A. (2014). Brand architecture management: The case of four tourist destinations in Catalonia. *Journal of Destination Marketing & Management*, 3(1), 48–58.

Elorrieta, B., Cerdan Schwitzguébel, A., & Torres-Delgado, A. (2022). From success to unrest: The social impacts of tourism in Barcelona. *International Journal of Tourism Cities*, 8(3), 675–702.

España. La Moncloa. (2020). *Maps of the de-escalation phases of COVID-19. The situation of each territory of Spain in the transition to the new normal.* Retrieved from: https://www.lamoncloa.gob.es/covid-19/Paginas/mapa-fases-desescalada.aspx

España. La Moncloa (2021). *Summary of COVID-19 initiatives.* Retrieved from: https://www.lamoncloa.gob.es/presidente/actividades/Documents/2021/290721_CumpliendoJunio21_AnexoII.pdf

España. Ministry of Health. (2020a). Early response plan in a COVID-19 pandemic control scenario. Retrieved from: https://www.sanidad.gob.es/profesionales/saludPublica/ccayes/alertasActual/nCov/documentos/COVID19_Plan_de_respuesta_temprana_escenario_control.pdf

España. Ministry of Health. (2020b). Plan for the transition to a new normal, 28 April 2020. Retrieved from: https://www.lamoncloa.gob.es/consejodeministros/resumenes/Documents/2020/PlanTransicionNuevaNormalidad.pdf

España. Ministry of Health. (2020c). State strategy against the second wave. Retrieved from: https://www.sanidad.gob.es/profesionales/saludPublica/ccayes/alertasActual/nCov/documentos/Estrategia_estatal_segunda_ola.pdf

España. Ministry of Health. (2022). Official information from the Ministry of Health concerning COVID-19. Retrieved from: https://www.sanidad.gob.es/profesionales/saludPublica/ccayes/alertasActual/nCov/home.htm

España. Real Decreto-Ley 21/2020 of 9 June. (2020). Urgent prevention, containment and coordination measures to address the health crisis caused by COVID-19. Retrieved from: https://www.boe.es/eli/es/rdl/2020/06/09/21/con

España. Real Decreto-Ley 463/2020 of 14 March. (2020). Declaration of the state of alert for the management of the health crisis caused by COVID-19. Retrieved from: https://www.boe.es/eli/es/rd/2020/03/14/463/con

España. Real Decreto-Ley 8/2020 of 17 March. (2020). Urgent extraordinary measures to address the economic and social impact of COVID-19. Retrieved from: https://www.boe.es/eli/es/rdl/2020/03/17/8/con

España. Real Decreto-Ley 962/2020 of 25 October. (2020). *Declaration of the state of alert to contain the spread of infections caused by SARS-CoV-2.* Retrieved from: https://www.boe.es/buscar/act.php?id=BOE-A-2020-12898#a5

EUROSTAT. (2021). Short-stay accommodation is offered via online collaborative economy platforms. Retrieved from: https://ec.europa.eu/eurostat/statistics-explained/index.php?title=Short-stay_accommodation_offered_via_online_collaborative_economy_platforms#In_2019.2C_more_than_1.5_million_tourists_per_night_slept_in_a_bed_booked_via_the_platforms

Garay-Tamajón, L., Lladós-Masllorens, J., Meseguer-Artola, A., & Morales-Pérez, S. (2022). Analysing the influence of short-term rental platforms on housing afforda-bility in global urban destination neighbourhoods. *Tourism and Hospitality Research*, 22(4), 444–461. https://doi.org/10.1177/14673 584211057568

Generalitat de Cataluña. (2007). LEY 15/2007, del 5 de diciembre, de la Agencia Catalana de Turismo. Retrieved from: https://dogc.gencat.cat/es/document-del-dogc/?documentId=446175

ICCA. (2020). ICCA statistics report country and city rankings. Retrieved from: www.iccaworld.org/dcps/doc.cfm?docid=2396

INE. (2022). Encuesta de ocupación hotelera. Instituto Nacional de Estadística. Retrieved from: www.ine.es/dynt3/inebase/index.htm?padre=238&capsel=238

Jutglá, E. D. (2019). Changing economic territories in the neighbourhoods of Poblenou and Sants in Barcelona—The effects of tourism (2005–2016). *Cuadernos de Turismo*, 43, 595–597.

Lalicic, L. & Önder, I. (2018). Residents' involvement in urban tourism planning: Opportunities from a smart city perspective. *Sustainability*, 10(6), 1852.

OTB. (2022a). Key figures. Retrieved from: https://www.observatoriturisme.barcelona/en/key-figures-2022

OTB. (2022b). Tourism activity report. Accommodation-Tourist demand-Mobility infrastructures. Retrieved from: https://www.observatoriturisme.barcelona/sites/default/files/230505_Càpsula%201%20IAOTB_0.pdf

OTB. (2022c). Tourism activity report. Evaluation-Job market-Sustainability. Retrieved from: https://www.observatoriturisme.barcelona/sites/default/files/230505_Càpsula%201%20IAOTB_0.pdf

OTB. (2022d). Tourism activity report. Tourist profile-Tourist expenses. Retrieved from: https://www.observatoriturisme.barcelona/sites/default/files/230321_Càpsula%202%20IAT.pdf

Romão, J., Domènech, A., & Nijkamp, P. (2021). Tourism in common: Policy flows and participatory management in the tourism council of Barcelona. *Urban Research & Practice*, 16(2), 1–24.

Vayá, E., Garcia, J. R., Murillo, J., Romaní, J., & Suriñach, J. (2018). Economic impact of cruise activity: The case of Barcelona. *Journal of Travel & Tourism Marketing*, 35(4), 479–492.

Wilson, J., Garay-Tamajon, L., & Morales-Perez, S. (2022). Politicising platform-mediated tourism rentals in the digital sphere: Airbnb in Madrid and Barcelona. *Journal of Sustainable Tourism*, 30(5), 1080–1101. https://doi.org/10.1080/09669582.2020.1866585

Wray, M. (2013). *Adopting and implementing a transactive approach to sustainable tour-ism planning: Translating theory into practice. Tourism governance* (pp. 205–228). Routledge.

2

TOURISTS' EXPERIENCES IN TIMES OF UNCERTAINTY

An analysis of eWOM before, during and after the COVID-19 pandemic

Ewa Wszendybył-Skulska and Adam Jezierski

Introduction

In recent times, the global hotel industry has developed rapidly. Yet, the unforeseen emergence of the COVID-19 pandemic severely affected this growth trajectory (Gretzel et al., 2020). Numerous countries and regions imposed quarantines, travel bans and other measures to curb the spread of the virus. This health crisis sped up shifts in global lifestyles, affecting how people work, communicate, learn and, notably, travel. Travel decisions have become intertwined with both internal and external factors, such as the perception of risk and official travel prohibitions. Furthermore, domestic restrictions, individual financial stability, available spending money, fluctuating costs and the apprehension of contracting the virus have profoundly reshaped tourist behaviour. So far, a few researchers had investigated how this change in tourists' habits and preferences influenced customers' satisfaction. Jezierski et al. (2022) showed that crisis-resistant tourists do not evaluate more harshly their hotel experiences because of the COVID-19 pandemic restriction. Yu et al. (2022) revealed that guest satisfaction is influenced by crisis response strategies employed by hotels. On the other hand Olivieri et al. (2023) reported an increase in the number of negative reviews during the first year of pandemic.

However, little research so far had focused on how electronic word of mouth (eWOM), generated by hotel visitors, have changed between the time before, during and after the pandemic (Olivieri et al., 2023). With COVID-19 restrictions lifted and hotel operations returning back to normal functioning it is crucial for hotel managers to better understand how customers assess various hotel features and how these evaluations impact their satisfaction (Xu et al., 2022). Similarly, most of the current literature regarding hospitality properties

DOI: 10.4324/9781032637778-3

had primarily evaluated the effects of COVID-19 on travel volume, revenue fluctuations and macro-perspectives (Sekar & Santhanam, 2022) showing the need for more research from micro-perspectives. Subsequently, the majority of eWOM research in regard to the COVID-19 pandemic is limited to data from TripAdvisor (Sekar & Santhanam, 2022) or Online Travel Agencies (OTAs) (Olivieri et al., 2023). More nuanced study might be acquired by scraping data from more often used sources – such as Google Reviews.

The given study aims to fill the mentioned research gap by utilising a mix-method approach and analysing both the evolution of review scores and the changes in topics being most frequently discussed in reviews' texts. The data for the study were acquired through the use of web-scrapping method on the sample of the Polish hotel market (which is one of the leading markets in Eastern Europe) (Cushman & Wakefield, 2023). The given research is one of the first to analyse eWOM originating before, during and after the COVID-19 pandemic. Additionally, focusing on eWOM trends changes allowed for providing hoteliers with key information regarding guests needs, which can be used to better adapt hotel services to resonate with the current market's preferences. Finally, our study provided an example of successful incorporation of data crawled from Google Reviews in tourism-related studies.

Literature review

Specificity of the hotel industry

The hotel industry constitutes a major part of the global economy. During the second decade of the 20th century, the global hospitality industry developed rapidly. According to Statista (2020), in 2018 there were 184,299 hotels around the world, which means that, compared to 2008, their number increased by almost 8.5%. In 2018, the retail value of the global hospitality industry was US$600.49 billion (AHLA, 2020). From an economic point of view, the hotel industry has so far been a significant source of a large number of jobs; exact numbers are difficult to estimate, which results from the specificity of the labour market, links with other industries, e.g., catering, and the seasonality of demand for hotel services. It is estimated that in 2019 the hotel industry worldwide provided 381 million jobs. In the United States alone, the hotel industry in 2019 provided 14 million jobs, including 2.3 million directly in hotels alone and generated US$300 billion in revenues (AHLA, 2020).

Similarly, in Europe the hotel industry was becoming a stronger pillar of the European economy, creating 1.6 million jobs between 2013 and 2016 and increasing the number of employees to 11.9 million. At the same time, the number of companies providing hotel services increased from 1.82 million to almost 2 million while the revenue rate increased by 20% from 507 to 607 billion euros (Hospitality Europe, 2017). By 2020, the European hotel industry directly employed 80% of the total EU tourism workforce and totalled more than 2 million enterprises (including approximately 200,000 hotels), of which 9 out of 10 are micro-enterprises with fewer than 10 employees (HOTREC, 2020).

The hotel industry, although subject to the general market forces, has many specific features. Firstly, it is characterised by high capital intensity as it implies a large value of fixed assets, long circulation (repayment) period and significant share of amortisation costs. High costs and a long-term investment process make hotel investment activities considered as highly risky projects (Rodríguez, 2002). Once a hotel facility is built, it cannot be enlarged or reduced excluding long-term activities. Similarly, the change of location is also difficult. Little flexibility and high sensitivity to both market conditions and competition result from stability of the potential because neither the hotel potential nor its utilisation program can be quickly adjusted to the constantly changing customer demand. These in turn impose certain limitations when it comes to human capital management. The hospitality industry offers fewer stable jobs than the rest of the employers. The percentage of people in temporary employment in this sector is more than twice as high (30%) than in the EU economy as a whole (14%). During the high season, the number of employees in hotel companies increases by an additional 10%. In addition, the high share of fixed costs forces hotel managers to focus primarily on the profitable side of their operations (Wszendybył-Skulska, 2012).

A characteristic feature of hotel services market is also the increased activity of private entrepreneurs and the domination of small and medium-sized enterprises in its composition (Marchante et al., 2005; Wszendybył-Skulska, 2012). The biggest problem, especially in highly industrialised countries, is the shortcomings related to the low status of the hotel industry, seasonality of demand or low wages. Employees of the hotel industry are the people with the lowest earnings in all industry comparisons (Casado-Díaz & Simón, 2016; Wszendybył-Skulska, 2012, 2013). Thus, the precarious status of employees of the hotel industry (Baum et al., 2016; Tapia & Alberti, 2019; Williamson & Harris, 2019) results mainly from the unsustainable and precarious employment and poor working conditions (which in turn results from low wages) (Baum et al., 2020; Robinson et al., 2020). This in turn makes them the hotel industry workers suffer the greatest consequences of social and market 'turbulences' and crises, as they are excluded from emergency solutions (Baum et al., 2020). The most typical decision made by hotel managers in crisis situations to protect themselves against the deterioration of current financial results is dismissing the employees.

Crisis impact on hotel industry

The hotel industry is highly susceptible to the consequences of disasters and external crises, from natural to man-made ones (Ritchie, 2004). With regard to the hotel industry, crisis phenomena can be classified into two groups, related to the impact on:

- The entire tourism economy, including entities from the hotel industry (i.e. natural disasters, climate changes, changes in the economic situation, industrial disasters, including transport disasters, social (internal) conflicts, military and political conflicts, migration crises, epidemics and pandemics, unfavourable conditions of the natural environment, including the condition of air and water purity (Łapko et al., 2020; Panasiuk, 2013).
- Directly to the hotel industry in two perspectives: mesoeconomic concerning the entire hotel industry and microeconomic concerning individual hotel enterprises.

Over the last decades, the hotel industry has faced many diverse crisis phenomena (Table 2.1). The industry suffered greatly as a result of the attacks of 11

TABLE 2.1 Crisis phenomena in the hotel and tourism sector in the last decades

September 11 terror attacks	
• sharp decline in travel demand.	
• suspension of airline services.	
• US hotel bookings declined by 20–50% in the first three months following 9/11.	
SARS	
• negatively impacted on the tourism economies of countries that are major tourist centres in Asia.	
South Korea	• average hotel occupancy rates across the industry fell by almost 30% between May and June 2003.
China	• decline in tourism traffic in 2003, compared to 2002, was approx. 9.4 million tourists.
	• decrease in incoming foreign tourism revenues of approx. US$30–50 billion.
	• decrease in the share of tourism in GDP by 25%.
	• decline in employment in the tourism economy by 22.8 million workers.
	• decline in hotel occupancy rates in Beijing by 23% in 2003 compared with those in 2002, and the occupancy rates of many four- and five-star hotels dropped to as low as 10%.
Hong Kong	• decline in the share of tourism in GDP by approx. 41%.
	• decline in employment in the tourism economy by 27,000 employees.
Singapore	• decline the share of tourism in GDP by approx. 43%.
	• decline in employment in the tourism economy by 17,500 employees.
Japan	• decline in revenues in the tourism economy was estimated at 47%.

Source: Dombey, 2004; Henderson & Ng, 2004; Kim et al., 2005; Kosová & Enz, 2012; Pine & Mckercher, 2004.

TABLE 2.2 Changes in the most important hotel profitability indicators

	OCC		*ADR*		*RevPAR*	
	2020/2019	*2021/2020*	*2020/2019*	*2021/2020*	*2020/2019*	*2021/2020*
Asia Pacific	−41.3	16.1	−20.4	5.1	−54.9	21.9
Americas	−35	26.6	−18.2	11.3	−53.8	64.3
Europa	−48.4	−3.8	−14	−1.7	−57.2	11.6
Middle East/ Africa	−36.4	16.1	−13.5	5.1	−44.8	21.75
Poland	−53.7	−6.9	−9.7	2.6	−58.2	−4.4

Source: *STR*, 2022.

September 2001 in New York, terrorist attacks in Europe in 2004 and 2005, the tsunami in Asia in 2004, the SARS epidemic in 2003 (mainly in the Far East), MERS in 2012 (mainly in South Korea and the Middle East) and the Ebola virus in 2014 in African countries (Maphanga & Henama, 2019; Mizrachi & Fuchs, 2016). These events caused significant, but short-term, restrictions in tourist traffic and thus resulted in a deterioration of the profitability of the hotel industry.

However, it is the COVID-19 pandemic that was both the most devastating thus far and possessed the most negative impact on the global economy. Strict regulations such as state of emergencies, stay-at-home orders and partial or complete border closings caused major shocks to overall economic activity. As with prior crises, hotel industry was the one of the most significantly affected by COVID-19 (Gretzel et al., 2020).

As a result of the mentioned restrictions, tourist traffic practically stopped at the beginning of 2020. In the following months it fluctuated in response to periods of relative normality and waves of intensification of the pandemic and related restrictions. The reduction in tourist traffic had a direct impact on the hotel industry worldwide. According to Smith Travel Research (STR) study, the values of the three most important hotel efficiency indicators (Room occupancy, ADR, RevPAR) were down in all parts of the world (Table 2.2).

The pandemic has also resulted in significant job losses in the hotel industry through furloughs and layoffs, which account for nearly 3.9 million jobs since the pandemic began (AHLA, 2020). One of the effects of limited tourist traffic were also qualitative changes related to the adaptation of hotels to the changing expectations of buyers and imposed formal restrictions. In a similar fashion, fears about long journeys resulted in an increase in domestic tourist traffic.

COVID-19 pandemic in Poland

The first cases of COVID-19 in Poland appeared in February 2020, and on 15 March 2020, the first restrictions were introduced. The detailed calendar of lockdowns and restrictions in Poland is presented in Table 2.3.

TABLE 2.3 Calendar of the COVID-19 pandemic in Poland

	Date	*Restrictions introduced for the tourism and catering industries*
1st lockdown	13 March 2020	Restrictions for restaurants to serve meals for take away only
		Closing gyms/fitness centres and swimming pools
		Putting a ban on organising social gatherings and parties
	23 March 2020	Restriction on movement except for living, health and professional purposes
	31 March 2020	Suspension of operations of hotel facilities and short-term rentals Staying on a beach and in green zones prohibited
	16 April 2020	Abolishment of restrictions on movement other than for living, health and professional purposes
		Lifting of the ban on staying on a beach or in green areas
	4 May 2020	Resumption of hotel operations, except for recreational venues such as swimming pools and fitness centres
	18 May 2020	18 May 2020 Lifting of the ban to serve meals for take away only, serving meals in the venues while maintaining sanitary regime
	30 May 2020	30 May 2020 Lifting of the ban on organising meetings and receptions up to 150 people
		Resumption of operations of swimming pools, sauna and steam rooms, and fitness centres
	8 August 2020	Introduction of division of regions into green zones (without additional restrictions), yellow zones (with partial additional restrictions) and red ones (with strong additional restrictions)
		In yellow zones, the number of people at special events is limited to 100, in red zones to 50;
		In yellow and red zones, restrictions on the operation of restaurants and bars until 9 o'clock pm and limiting the operation of swimming pools and fitness centres
2nd lockdown	23 October 2020	The whole of Poland is declared a red zone
		Putting a ban on organising special events
		Suspension of in-house restaurant operations
	7 November 2020	Hotel services available to those on business trips only
	17 December 2020	Further restriction of the availability of hotel services limited to people on the list of permitted business trips only
	12 February 2021	Re-opening of hotels to all guests (while maintaining the sanitary regime)
		Re-opening of swimming pools
		Restrictions on gyms and restaurants remain unchanged

(Continued)

TABLE 2.3 (Continued)

	Date	Restrictions introduced for the tourism and catering industries
	20 March 2021	Limitation of hotel operations to providing accommodation to people on the list of permitted business trips only Closing of swimming pools
3rd lockdown	3 May 2021	Re-opening of hotels to all guests (while maintaining the sanitary regime, limiting the facility's occupancy to 50%, restaurants and wellness zones closed)
	15 May 2021	Allowing restaurant guests to be served outside (e.g., restaurant gardens) while maintaining the sanitary regime Possibility of organising special events for up to 25 customers outside of a venue
	29 May 2021	Allowing restaurant guests to be served inside while maintaining the sanitary regime Possibility of organising special events for up to 50 customers inside of a venue
	6 June 2021	Possibility of organising special events for up to 150 customers outside of a venue The limits do not apply to the fully vaccinated
	26 June 2021	Increasing the occupancy limit for hotels and restaurants to 75% of occupied rooms, the limit does not apply to organised groups of children and adolescents under 12 years of age The limits do not apply to the fully vaccinated
4th lockdown	15 December 2021	Restaurants, bars and hotels – max. 30% occupancy by the unvaccinated The limits do not apply to the fully vaccinated
	1 March 2022	Lifting the limits, hotels can make 100% of their beds available again

16 May 2022 end of the COVID-19 pandemic in Poland

Source: (Koronawirus: Informacje i Zalecenia, 2023).

It is worth mentioning that many hotels in Poland during the COVID-19 pandemic decided to seize the opportunity to direct their offer to doctors, paramedics and health care workers (becoming the so called 'Hotels for Medics'). That strategy was particularly suitable for hotels located in large agglomerations with many hospitals and medical centres that treated patients infected with the COVID-19 virus (Podhorodecka & Bąk-Filipek, 2022). The medical staff of those facilities could take advantage of free accommodation in hotels without having to return home and expose their families to the risk of infections. The costs of stay of the medical staff were covered by the Foundation of the Polish Hotel Holding. Therefore, these hotels were able to operate normally and generate profits during the pandemic.

The COVID-19 pandemic caused people to experience negative emotions, fear and anxiety about infection, which directly influenced travel intentions and purchasing decisions, especially in the field of hotel services (Gretzel et al., 2020). During the pandemic, decisions about the selection and purchase of hotel services were influenced by many factors, including but not limited to destinations' environment and preferences (Jang et al., 2021). COVID-19 has impacted people's choices about where to visit when travelling. The pandemic also had a significant impact on psychological perceptions and behavioural responses. Therefore, this study focused on differences in the assessment of basic hotel attributes before and after the pandemic.

Electronic word of mouth

Electronic word of mouth (usually called eWOM) is defined by Rosario et al. (2020, p. 247) as a 'consumer-generated, consumption-related communication that employs digital tools and is directed primarily to other consumers'. This definition was conceptualised based on the extensive literature research and showcases the key elements of the construct:

- eWOM is created by consumers as opposed to professionals or businesses. Reviews created by journalists or paid influencers cannot be considered consumer-generated (Naab & Sehl, 2017).
- eWOM need to relate to either past, present or future purchase and consumption of either products or services (Yen & Tang, 2019).
- eWOM needs to be published in online media (Mishra & Satish, 2016).
- eWOM, as a form of traditional Word of Mouth (WOM), should be orienteered at sharing information towards other potential consumers, and as such be easily available for general public to interact with (Rosario et al., 2020).

Huete-Alcocer (2017) underlines that eWOM differs from traditional WOM in being less trustworthy, but at the same time more accessible to consumers. Subsequently, eWOM has shifted the equilibrium and transformed the communication landscape between manufacturers, retailers, their brands and consumers (Mishra & Satish, 2016) by allowing for more direct and unstructured conversations between the business and customers (Bhaiswar, 2020). The importance of eWOM origins from its high impact of purchase decisions. In the context of tourism industry, eWOM has proved to be one of the most important and trustworthy source of information, universally used by tourist of all levels of income and experience in travelling (Ladhari & Michaud, 2015; Zaman et al., 2016). Many authors also agree that positive eWOM can be a significant factor in empowering booking intensions (Lee et al., 2021; Serra-Cantallops et al., 2018; Zaman et al., 2016).

eWOM is also an important aspect from the perspective of hotel businesses. Anagnostopoulou et al. (2020) established that positive eWOM had a positive influence on the profits of hotels in Great Britain while the negative eWOM had an opposite effect. At the same time, hotel properties that actively manage reviews connected with their hotels observe positive benefits. Kim et al. (2015) showcased that replying to negative reviews posted on the Internet has led to the increase in hotels' Average Daily Rates (ADR) and Revenue Per Available Room (RevPAR). Similarly, Xie et al. (2017) observed that increasing the number of managerial replies to eWOM was directly and positively correlated with properties RevPAR. eWOM was also seen as a factor that can be used to build customers' loyalty (Serra-Cantallops et al., 2018). It is worth noticing that eWOM can assume different forms. Mishra and Satish (2016) showcased that it can function either text (in the form of reviews, opinions, or blog posts), reactions (such as shares or likes on social media posts) or numeric values (grade system attached to opinions on certain review websites). In the context of tourism industry, both researchers and tourists predominantly use review sites like TripAdvisor and Google Reviews, online travel agencies like Booking.com and social media channels such as Facebook and Instagram (Filieri, 2016).

eWOM in times of COVID-19

Ladhari and Michaud (2015) specified eWOM to be especially useful in situations of uncertainty or lack of proper information regarding the hotel properties. This is why, eWOM become an even more important aspect for tourists, as the time of COVID-19 pandemic was a period of high uncertainty in almost all aspects (Jezierski et al., 2022). Nilashi et al. (2022) were investigating eWOM impact on tourists from different segments, travelling during the time of COVID-19 pandemic. The authors showed that while eWOM remained an important information source for travellers, its credibility, perceived usefulness and helpfulness were moderated by the travellers' perspective of risk and their e-trust. The results of the research also showed that gender and travel experience also moderate the association between e-trust and perceived risk, and between perceived risk and intention to travel (Nilashi et al., 2022).

COVID-19 has also influenced the characteristics of eWOM produced by the tourists. Jezierski et al. (2022) found that reviews which mention the pandemic were typically longer and considered more helpful by other travellers. It suggested that there was a need for COVID-related information by potential customers looking for hotel destinations. At the same time, reviews relating COVID-specific aspects of the stay were generally of the same score as other reviews, implying that pandemic become one of typical issues for travellers in Poland (Jezierski et al., 2022). Similar results were obtained by Mehta et al. (2021), who showed that hotels located in Europe have managed to maintain their levels of service in all periods of the investigation. It stood as an opposition to properties in Africa and Australia, where even high-end hotels were not able to handle low number of tourists and were showed to be unequipped to adapt to new situation (Mehta et al., 2021). It is worth noticing that both of the mentioned studies have reported a general decline in volume of reviews during the analysed pandemic times.

Hu et al. (2021) went on to compare eWOM between two timeframes – 'ongoing pandemic' and 'recovering period' – and reported a change in reviews' different attributes importance. Universally, aspects connected with hygienic aspects of the stay, such as cleanliness or level of service increased in importance both during and after the most severe time of the pandemic. At the same time, it seemed that tourists were more understanding towards lacks in quality in different fields – a finding in line with Jezierski et al. (2022). Because of this, there was a significant decrease in the importance of breakfast, location and surroundings (Hu et al., 2021). Similarly, the authors reported that while issues connected with food services were often mentioned in the reviews, the tourists showed a lot of understanding in that aspect and it did not impact the final opinion score. An attempt of comparing pre- and post-COVID-19 reviews was also undertaken by Xu et al. (2022). Their research showed that while the importance of location did not diminish in pandemic times, its context changed to emphasise other factors such as being close to a hospital. A similar change was observed in terms of staff context, which was now more related to the theme of epidemic prevention. At the same time, Xu et al. (2022) observed, in line with Hu et al. (2021), that catering services remained being mentioned in reviews, but their significance for assessing quality of service dropped.

A shift in key subjects mentioned by tourists in reviews was also observed by Olivieri et al. (2023). Out of the four main topic groups (travel experience, travel organisation, travel expenses and costs, recommendations) only the travel experience was identified in reviews published during the pandemic times. Instead, three other topic groups emerged – refund opportunity, travellers satisfaction and company support. Across all of these groups a separate issue of the easiness of contacting hotels was prevalent in all kind of reviews (Olivieri et al., 2023). When it comes to the sentiment of the reviews, it was shown that while positive reviews volume remained stable, there was a significant increase in the number of negative reviews. Such a results would suggest that in the times of uncertainty and shift in priorities, hotel properties are not adjusting their policies and packages in line with tourist demands. Somewhat opposite results were obtained by Sekar and Santhanam (2022) who showcased a clear dominance of positive reviews when analysing hotel sector during COVID-19 pandemic. Moreover, the authors observed that while for hotels in all types of location (metro/urban/rural) the reviews were predominantly positive, the highest share of positive eWOM was seen in terms of rural properties (suggesting its location being an important aspect during the pandemic times).

Ensuring comprehensive safety and hygiene measures for customers and prioritising their well-being throughout their stay during the pandemic was a new way for hotels to provide quality service (Sekar & Santhanam, 2022).

Impact of COVID-19 crisis on eWOM

Despite several studies confirming a shift in both the sentiment and the focus of the hotels' reviews in time of COVID-19 (Hu et al., 2021; Mehta et al., 2021; Olivieri et al., 2023; J. Xu et al., 2022), the question remains if such an change is reversible or becomes a 'new normal' for post-pandemic tourists. In such a sense Ho et al. (2021) called for future studies that would 'uncover further differences among "Before COVID-19," "With COVID- 19," and "After COVID-19" (pre/in/post) periods'. Similarly, Jezierski et al. (2022) also pointed out the need of repeating analysis of COVID-related eWOM in different time periods. Another aspect that needs further investigation is the average sentiment of reviews during the COVID-19 period as conflicting results were obtained in the past (Olivieri et al., 2023; Sekar & Santhanam, 2022). As such, it is hypothesised that:

> H1: The COVID-19 pandemic had influenced the general score of hotels' eWOM

The aspect of differences in travellers' approach to hotels' location during the COVID-19 pandemic was showcased in review analysis by both Xu et al. (2022) and Sekar and Santhanam (2022). However, also in times before the pandemic hotels located in rural and urban locations were observing different levels of satisfaction from their customers. For example, Xu and Li (2016) pointed out hotels' location as an aspect frequently brought up in positive guests reviews. Therefore, it is hypothesised as follows:

> H1a: The COVID-19 pandemic had influenced the location related score of hotel reviews

Subsequently, the issue of the level of staff service and guests' attitude towards them was brought up frequently in the context of COVID-19 pandemic (Mehta et al., 2021; Xu et al., 2022). On the other hand, already in pre-COVID times Jezierski (2022) proved that the satisfaction from staff service is determined by various factors such as hotels' age, star rating or number of rooms. Thus, it is hypothesised that:

> H1b: The COVID-19 pandemic had influenced the staff related score of hotels' eWOM

COVID-19 period forced hotel to introduce new safety standards (sometimes enforced by governments, sometimes introduced by hotels own violation), which lead to many properties emphasising cleanliness in their promotional campaigns (Shin & Kang, 2020). The importance of this aspect was universally noted as critical both during the pre-pandemic times (Jezierski, 2022) and during it (Liu & Hu, 2022; Mehta et al., 2021). Xu et al. (2022) connected the idea of cleanliness to the room aspect and suggested that hotel managers during the COVID-19 pandemic ought to focus more of room conditions in terms. Because of it, it is hypothesised that:

H1c: The COVID-19 pandemic had influenced the room related score of hotels' eWOM

Various studies had also pointed out a general shift in the topics and themes observed in the pandemic time reviews (Mehta et al., 2021; Olivieri et al., 2023). It is still uncertain if such a change of focus will be reversed after the pandemic ended. Because of that, Olivieri et al. (2023) called for repeated studies which would verify the already observed topic shifts in more generally available review websites as well as would check if there is a new change in regard to post-pandemic eWOM. In a similar way, Jezierski et al. (2022) called for a more versatile research in terms in the context of COVID-19-related eWOM. Therefore it is hypothesised that:

H2: The COVID-19 pandemic had shifted the key themes seen in hotels' eWOM

Methodology

Data collection

We collected data from Google profiles of all registered Polish hotels which possessed such a profile. Google Reviews were chosen as a source for the dataset, as it is the largest website that records customers' review over time without deleting past records – see Table 2.4 based on data from SimilarWeb statistics (2023).

TABLE 2.4 Review websites statistics

Website	Total visits (billions)	Monthly visits (millions)	Global rank
TripAdvisor	0.5	183.8	#228
Booking.com	1.9	649.7	#46
Google.com	255.4	85,160	#1

Booking.com was excluded as a possible data source as it removes reviews that are older than three years, and this process would make it impossible to compare reviews in pre-pandemic times. Google Reviews already served as a data source in eWOM-related research, e.g., Mathews et al. (2021). A total of 528 361 reviews were collected in August 2023 originating from 2,289 hotel properties. Octoparse software (version 8.0) was used to conduct the web-scrapping. Such a software was used as it provides the best options to counter 'lazy-load' (loading more reviews only when the end of the page is reached) nature of Google Reviews website. The following information was downloaded for each of the reviews: review's author (*r_author*), review's text (*r_text*), review's score (*r_score*) and review's date of publication (*r_date*). Additionally, if such information was available, additional information was acquired such as review's score of location (*r_location*), review's score of staff (*r_staff*) and review's score of room (*r_room*). The obtained data were then divided into three samples, based on the reviews' publication dates:

- pre-pandemic reviews – reviews posted before 2020.
- pandemic time reviews – reviews posted in the years of 2020 and 2021.
- post-pandemic reviews – reviews posted after 2021.

The showed above division of the research period was used, as the first COVID-related changes in Poland were recorded in February 2020 and most of the COVID-related restrictions were lifted in Poland in the first quarter of 2022.

Data analysis

The data analysis process was divided into two parts. Firstly, the normality of the data was verified using the Kołmogorov-Smirnov test (Shapiro-Wilk test was not used due to the large sample size). As the normality assumption was violated, following the works of Gerdt et al. (2019), to evaluate the pandemic's impact on the reviews score, Kruskal-Wallis test was used. Non-parametric test was used in place of analysis of variance, as datasets related to eWOM (especially reviews) tend to possess either a bimodal or a 'J-shaped' distribution (Zervas et al., 2021). Therefore, a non-parametric alternative to the one-way ANOVA was selected. Reviews' general score as well as reviews' score of staff, room and location was compared between the mentioned timeframes. If Kruskal-Wallis test showed significant differences, then as a post hoc test a Dunn test was used. Moreover, Kruskal-Wallis test was done separately for each subsample created for hotels' star category which acted as moderating variables for the relationship.

Secondly, a text analysis of the eWOM was conducted. As the first step, a sub-sample of the reviews containing 52,825 texts with their scores were selected for further analysis. The created sub-sample was created with Statista and was checked to contain the same distribution among the scores and star categories as the whole sample. Then, the data were uploaded into the MAXQDA software and a separate world count for each of the periods (pre-, during and post-pandemic) was conducted using one of the options from MaxDicio group. Both lemmatisation option and manual check of the results were used to group words of similar meaning.

Results

The descriptive results of the qualitative part are presented in Table 2.5. Out of the 523,957 analysed reviews 34% were published before, 22% were written during and 44% after the pandemic. There were, however, significantly less scores for room, staff and location as the possibility to evaluate this aspect was introduced by Google only a short time before 2022. When it comes to the distribution of the reviews between different star categories, it is reflecting the distribution of hotels in Poland (the dominance of 3* properties and low number of 2* properties). A surprising outcome was a fact that across all the sample the median value of the reviews was '5' – the highest.

Kruskal-Wallis test results are presented in Table 2.6. As can be seen, there were significant differences identified for all type of scores in all subsections of the sample. Worth noticing is the fact that all tests were significant at $p < 0.001$ apart from the score of the room ($p < 0.01$) and the score of the location ($p < 0.05$) – both in 2* hotels subsection. Subsequently, a visual analysis of shapes of distribution in each subsection was conducted and showed to be of a similar shape (Annex 1). Therefore, following the works of Nwobi and Akanno (2021), the existence of stochastic dominance between populations allows for comparison of means between populations.

Dunn tests results for analysing the general score of the reviews can be found in Table 2.7. As can be seen, there was an increase in average scores both between pre- and during the pandemic as well as during the pandemic and post times. It would suggest that despite the coronavirus restrictions and changes in the services profile, Polish hoteliers were able to maintain a high standard of services offered. However, while statistically significant, the mean differences are of small power. Across the whole sample the difference between pre- and post-pandemic times was about 0.15 in average review score. The highest differences were observed for hotels with lower star rating (especially 1* properties). Such a phenomenon might be explained by the fact that customers have lower expectations for low star properties, and because of that, it is easier to exceed them. Moreover, the pandemic restrictions limited the offering of certain services in high-end properties, making the product offer in all star categories similar. Thus, based on the findings, Hypothesis 1 was deemed supported.

Analysing the impact of the pandemic on the scores of room, there was an average increase of 0.42 in the room review score before and after the

TABLE 2.5 Descriptive results

Star category		r_score			r_room			r_staff			r_location		
		Pre	Pand	Post	Pre	Pand	Post	Pre	Pand	Post	Pre	Pand	Post
1	N	28172	19149	43754	671	3625	28291	639	3446	27616	565	3127	25567
	Mean	4.18	4.31	4.35	3.95	4.18	4.42	3.94	4.24	4.52	4.28	4.47	4.62
	Median	5.00	5.00	5.00	5.00	5.00	5.00	5.00	5.00	5.00	5.00	5.00	5.00
	Std. Deviation	1.17	1.16	1.16	1.49	1.34	1.07	1.50	1.31	1.03	1.21	1.06	0.83
2	N	4412	2409	4079	84	423	2541	80	402	2484	71	368	2319
	Mean	4.11	4.23	4.26	3.55	4.10	4.27	3.59	4.19	4.49	3.99	4.42	4.48
	Median	5.00	5.00	5.00	5.00	5.00	5.00	5.00	5.00	5.00	5.00	5.00	5.00
	Std. Deviation	1.18	1.21	1.23	1.75	1.41	1.19	1.77	1.43	1.08	1.39	1.16	1.02
3	N	99849	61441	107119	2265	11957	69941	2145	11481	68243	1933	10471	63190
	Mean	4.13	4.23	4.28	3.89	4.08	4.28	3.97	4.17	4.47	4.30	4.45	4.57
	Median	5.00	5.00	5.00	5.00	5.00	5.00	5.00	5.00	5.00	5.00	5.00	5.00
	Std. Deviation	1.17	1.19	1.21	1.49	1.39	1.17	1.48	1.37	1.09	1.18	1.07	0.90
4	N	32139	21766	47971	919	4492	31565	865	4297	30853	784	3884	28577
	Mean	4.15	4.24	4.30	3.91	4.07	4.33	3.96	4.15	4.49	4.26	4.42	4.58
	Median	5.00	5.00	5.00	5.00	5.00	5.00	5.00	5.00	5.00	5.00	5.00	5.00
	Std. Deviation	1.17	1.19	1.20	1.47	1.39	1.15	1.44	1.37	1.07	1.16	1.08	0.88
5	N	15459	10935	25303	402	2295	16450	376	2188	15948	347	1968	14557
	Mean	4.23	4.26	4.32	4.06	4.17	4.40	4.15	4.18	4.50	4.43	4.55	4.64
	Median	5.00	5.00	5.00	5.00	5.00	5.00	5.00	5.00	5.00	5.00	5.00	5.00
	Std. Deviation	1.12	1.19	1.20	1.41	1.33	1.10	1.41	1.34	1.07	1.07	0.96	0.82

TABLE 2.6 Kruskal-Wallis test results

Kruskal-Wallis test	r_score			r_room			r_staff			r_location		
	p	DoF	X2	p	DoF	X2	p	DoF	X2	p	DoF	X2
All	<.001	2	4809.77	<.001	2	568.79	<.001	2	1418.5	<.001	2	406.03
1*	<.001	2	896.41	<.001	2	106.42	<.001	2	251.11	<.001	2	77.26
2*	<.001	2	111.54	.004	2	11.108	<.001	2	31.13	.011	2	9.09
3*	<.001	2	2381.8	<.001	2	205.10	<.001	2	690.17	<.001	2	155.36
4*	<.001	2	917.46	<.001	2	170.9	<.001	2	392.99	<.001	2	144.15
5*	<.001	2	291.23	<.001	2	60.87	<.001	2	146.18	<.001	2	30.65

TABLE 2.7 Dunn test results for r_score

Star category		Test statistic	Std. error	Std. test statistic	Sig.	Adj. sig.[a]	Mean difference
All	Pre-Pand	−18506.194	500.918	−36.945	0.000	0.000	−0.10
	Pre-Post	−28972.262	418.929	−69.158	0.000	0.000	−0.15
	Pand-Post	−10466.067	479.639	−21.821	0.000	0.000	−0.05
1	Pre-Pand	−3677.253	211.595	−17.379	0.000	0.000	−0.13
	Pre-Post	−5056.746	172.564	−29.304	0.000	0.000	−0.17
	Pand-Post	−1379.494	195.756	−7.047	0.000	0.000	−0.05
2	Pre-Pand	−463.632	70.735	−6.554	0.000	0.000	−0.13
	Pre-Post	−621.527	60.661	−10.246	0.000	0.000	−0.16
	Pand-Post	−157.896	71.750	−2.201	0.028	0.083	−0.03
3	Pre-Pand	−9483.683	350.913	−27.026	0.000	0.000	−0.10
	Pre-Post	−14575.026	301.066	−48.411	0.000	0.000	−0.15
	Pand-Post	−5091.343	346.377	−14.699	0.000	0.000	−0.05
4	Pre-Pand	−3265.230	225.230	−14.497	0.000	0.000	−0.09
	Pre-Post	−5601.042	184.936	−30.286	0.000	0.000	−0.15
	Pand-Post	−2335.812	209.709	−11.138	0.000	0.000	−0.07
5	Pre-Pand	−1112.081	160.331	−6.936	0.000	0.000	−0.03
	Pre-Post	−2220.811	130.973	−16.956	0.000	0.000	−0.09
	Pand-Post	−1108.730	146.857	−7.550	0.000	0.000	−0.06

pandemic (as seen in Table 2.8). Least significant changes were reported for 2* properties (especially with adjusted significance) and partially for 5* properties. For the other star categories the difference in room score before and after the pandemic was quite similar and close to the value for the whole sample. It can be suspected, that during the times of restrictions and free manpower hotels might have decided to renovate some of the rooms which was subsequently seen in the review scores. As such, the Hypothesis 1a was deemed supported.

TABLE 2.8 Dunn test results for r_room

Star category		Test statistic	Std. error	Std. test statistic	Sig.	Adj. sig.[a]	Mean difference
All	Pre-Pand	−6004.236	717.224	−8.371	0.000	0.000	−0.19
	Pre-Post	−11432.638	666.686	−17.148	0.000	0.000	−0.42
	Pand-Post	−5428.402	308.508	−17.596	0.000	0.000	−0.22
1	Pre-Pand	−1226.214	325.342	−3.769	0.000	0.000	−0.23
	Pre-Post	−2251.976	302.379	−7.448	0.000	0.000	−0.47
	Pand-Post	−1025.762	136.568	−7.511	0.000	0.000	−0.24
2	Pre-Pand	−223.468	90.797	−2.461	0.014	0.042	−0.55
	Pre-Post	−269.536	84.294	−3.198	0.001	0.004	−0.72
	Pand-Post	−46.067	39.916	−1.154	0.248	0.745	−0.17
3	Pre-Pand	−3024.931	481.365	−6.284	0.000	0.000	−0.20
	Pre-Post	−5029.750	448.463	−11.216	0.000	0.000	−0.39
	Pand-Post	−2004.819	207.874	−9.644	0.000	0.000	−0.20
4	Pre-Pand	−1010.637	328.257	−3.079	0.002	0.006	−0.16
	Pre-Post	−2533.905	303.408	−8.351	0.000	0.000	−0.42
	Pand-Post	−1523.268	144.586	−10.535	0.000	0.000	−0.26
5	Pre-Pand	−365.823	245.924	−1.488	0.137	0.411	−0.11
	Pre-Post	−1034.977	229.612	−4.508	0.000	0.000	−0.34
	Pand-Post	−669.154	101.352	−6.602	0.000	0.000	−0.23

When it comes to the analysis of the scores assigned to the quality of staff service, the change during the pandemic was both significant and high – averaging around the increase of 0.51 across the whole sample (which can be seen in Table 2.9). Interestingly, the only non-significant change was reported for 5* properties, which might suggest that the level of service in this kind of properties is maintained in and out the times of crises. The change before and after the pandemic was also the lowest one observed. A sharp increase in value of the staff score was reported for 2* properties as well as 3* properties. What is worth mentioning is the fact that all types of properties observed an increase in staff evaluation, suggesting that the pandemic did not negatively affect guests' perception of the work of hotel employees. Nevertheless, based on the findings Hypothesis 1b was deemed supported.

The results for the Dunn test for the score of the location are presented in Table 2.10. The changes in the location scores in the pre- and post-pandemic periods were significant for the whole sample, but the power of the change was moderate. It averaged less than 0.3 for the whole sample. The strongest, but statistically least significant, change was reported for 2* properties. The change was more impactful for 3* and 4* hotels, for which the location score has improved by 0.27 and 0.32 respectively. Interestingly, the increase of the scores between pre- and during pandemic and between during and post-pandemic

TABLE 2.9 Dunn test results for r_staff

Star category		Test statistic	Std. error	Std. test statistic	Sig.	Adj. sig.[a]	Mean difference
All	Pre-Pand	−6431.493	654.880	−9.821	0.000	0.000	−0.21
	Pre-Post	−15206.816	609.089	−24.967	0.000	0.000	−0.51
	Pand-Post	−8775.323	279.819	−31.361	0.000	0.000	−0.31
1	Pre-Pand	−1582.189	300.721	−5.261	0.000	0.000	−0.30
	Pre-Post	−3084.157	279.379	−11.039	0.000	0.000	−0.58
	Pand-Post	−1501.968	126.140	−11.907	0.000	0.000	−0.28
2	Pre-Pand	−266.396	80.861	−3.295	0.001	0.003	−0.60
	Pre-Post	−370.235	75.026	−4.935	0.000	0.000	−0.90
	Pand-Post	−103.839	35.508	−2.924	0.003	0.010	−0.30
3	Pre-Pand	−2916.472	436.188	−6.686	0.000	0.000	−0.20
	Pre-Post	−6877.917	406.630	−16.914	0.000	0.000	−0.50
	Pand-Post	−3961.445	187.055	−21.178	0.000	0.000	−0.30
4	Pre-Pand	−1561.332	300.411	−5.197	0.000	0.000	−0.19
	Pre-Post	−3618.761	277.904	−13.022	0.000	0.000	−0.53
	Pand-Post	−2057.428	131.259	−15.675	0.000	0.000	−0.34
5	Pre-Pand	−992.112	213.371	−4.650	0.000	0.000	−0.03
	Pre-Post	−1061.938	93.232	−11.390	0.000	0.000	−0.35
	Pand-Post	69.826	228.302	0.306	0.760	1.000	−0.32

TABLE 2.10 Dunn test results for r_location

Star category		Test statistic	Std. error	Std. test statistic	Sig.	Adj. sig.[a]	Mean difference
All	Pre-Pand	−5728.677	622.884	−9.197	0.000	0.000	−0.16
	Pre-Post	−9236.917	579.502	−15.939	0.000	0.000	−0.29
	Pand-Post	−3508.239	264.935	−13.242	0.000	0.000	−0.13
1	Pre-Pand	−1868.243	268.890	−6.948	0.000	0.000	−0.19
	Pre-Post	−1183.821	288.999	−4.096	0.000	0.000	−0.34
	Pand-Post	−684.421	119.769	−5.715	0.000	0.000	−0.15
2	Pre-Pand	−225.707	76.081	−2.967	0.003	0.009	−0.44
	Pre-Post	−237.832	81.854	−2.906	0.004	0.011	−0.49
	Pand-Post	12.125	35.434	0.342	0.732	1.000	−0.05
3	Pre−Pand	−3874.610	387.231	−10.006	0.000	0.000	−0.15
	Pre-Post	−2447.198	415.158	−5.895	0.000	0.000	−0.27
	Pand-Post	−1427.412	176.947	−8.067	0.000	0.000	−0.12
4	Pre-Pand	−2403.233	266.720	−9.010	0.000	0.000	−0.16
	Pre-Post	−1340.892	288.473	−4.648	0.000	0.000	−0.32
	Pand-Post	−1062.340	126.000	−8.431	0.000	0.000	−0.16
5	Pre-Pand	−791.593	191.092	−4.142	0.000	0.000	−0.12
	Pre-Post	−463.290	204.828	−2.262	0.024	0.071	−0.22
	Pand-Post	−328.303	84.492	−3.886	0.000	0.000	−0.10

was similar for all category of hotels. Based on the findings, Hypothesis 1c was deemed supported.

The results of the content analysis, presented in Table 2.11, showed that the first 15 words that appear most frequently in reviews have not changed substantially when comparing periods before, during and after the pandemic. One of the noticeable changes concerns the word *restaurant*, which appeared in the list of 15 most frequently mentioned words only before the pandemic. This is probably related to the large restrictions imposed on hotels during the pandemic, making restaurant unavailable to guests during and between lockdowns. During the lockdown, hotels that still could operate were obliged to serve meals in guests' rooms. This is probably why, during and after the pandemic, the word that appeared among the 15 most frequently mentioned words in reviews were *breakfast* or *food*, as a basic meal most often offered by hotels in the price of accommodation as well as most easy to serve in the hotel room for properties that did not offered such services pre-pandemic.

The second word that was in the list of 15 most common words in reviews before the pandemic and does not appear in the lists during and after the pandemic is *price*. That reduced frequency can be explained by the fact that a hotel stay, especially during the pandemic, was treated as a luxury 'good' to which access was severely limited. Thus, guests appreciated the fact that they could stay in hotels at all and the price they paid was no longer as important to them as before the pandemic. The list of words most frequently mentioned in reviews after the pandemic shows that price is still not often mentioned.

A word that started to appear quite often in reviews after the pandemic is *location*, which was referenced in over 9% of all reviews, suggesting a possible shift in its importance for customers. The change in the ranking of the most frequently mentioned words is also visible in the case of *cleanliness*. The increase in its frequency in reviews seems to be a natural process of change as during the pandemic, people paid more attention to cleanliness than before. *Hotel* remains the most frequently mentioned word in reviews before, during and after the pandemic, which is of no surprise. Another word that has not changed its place in the ranking of the most frequently mentioned in reviews is *room*. Before, during and after the pandemic, this word was the fifth most frequently mentioned in reviews. *Service*, *a lot* and *very* also remained being most frequently mentioned in reviews in all analysed periods.

A comparative analysis of the most frequently mentioned reviews before and after the pandemic, despite showing certain shifts in rankings, proved that the key elements of hotel reviews remained unchanged. Based on this H2 was deemed unsupported.

TABLE 2.11 Content analysis results

Before the COVID-19 pandemic					During the COVID-19 pandemic					After the COVID-19 pandemic				
Word	Total frequency	% Total frequency	No. of reviews with word	% of reviews with word	Word	Total frequency	% Total frequency	No. of reviews with word	% of reviews with word	Word	Total frequency	% Total frequency	No. of reviews with word	% of reviews with word
hotel	6249	2.53	5293	30.17	hotel	4549	2.48	3851	32.83	hotel	11150	2.63	9274	40.12
a lot	4662	1.89	4450	24.76	very	4521	2.46	3411	29.34	a lot	8520	2.01	8068	34.89
very	5496	2.22	4282	23.82	service	3530	1.92	3437	29.30	very	10097	2.38	7467	32.29
service	4039	1.64	3957	22.02	a lot	3278	1.79	3113	26.54	service	7185	1.70	6978	30.18
room	4505	1.82	3909	21.74	room	3360	1.83	2886	24.60	room	7702	1.83	6698	28.96
good	4064	1.65	3664	20.38	recommend	2634	1.43	2582	22.44	information	6567	1.54	6515	28.18
information	3569	1.45	3537	19.67	nice	2668	1.46	2580	21.99	recommend	5388	1.20	5241	22.66
recommend	3305	1.34	3245	18.05	information	2538	1.39	2517	21.46	nice	5176	1.22	4981	21.54
food	3230	1.32	3138	17.46	good	2470	1.35	2245	19.14	good	4985	1.18	4513	19.52
nice	3151	1.28	3063	17.20	cleanliness	2248	1.83	2188	18.66	cleanliness	4595	1.08	4456	19.27
place	2946	1.19	2718	15.12	food	2186	1.19	2139	18.23	food	4396	1.04	4294	18.58
cleanliness	2802	1.14	2714	15.10	breakfast	1871	1.03	1831	15.61	breakfast	4273	1.00	4182	18.08
price	1479	0.60	1402	7.80	place	2111	1.15	1928	16.44	place	4169	0.97	3797	16.41
super	1411	0.57	1315	7.32	tasty	1169	0.63	1151	9.83	super	2600	0.61	2379	10.30
restaurant	1354	0.54	1297	7.22	super	1209	0.66	1116	9.52	localisation	2101	0.50	2089	9.03

Discussion

The COVID-19 pandemic ushered in an era of unprecedented change for the hotel industry, compelling it to re-evaluate and adapt its operational strategies. Concurrently, the pandemic reshaped the needs and expectations of travellers. As a direct reflection of these evolving dynamics, eWOM generated by tourists witnessed noticeable shifts in tone (but not in content) before, during, and after the pandemic.

Presented results are supporting the findings of Sekar and Santhanam (2022) as well as these of Mehta et al. (2021) by showcasing that the level of service in European hotels did not decrease during pandemic. The general increase in scores during and after the pandemic would suggest that hotel owners have used the times of enforced closures constructively to work on improving their products. At the same time, the results suggests both that Polish hoteliers were adaptive to the needs of their customers during the pandemic times and were able to shift back to 'normal' services in the times when restrictions were lifted. It stands in contrast to the findings of Jezierski et al. (2022), who did not find score differences between COVID-related and COVID-not-related reviews. However, such a discrepancy might result from different periods being investigated (only the time of 1st Opening during the pandemic). Similarly, our findings were opposite to these of Olivieri et al. (2023), who identified an increase in negative eWOM during pandemic. However, their data were collected from one specific OTA, which level of services during the cancellations time might have skewed the data.

Moreover, the presented results show that despite the employment in hotels being limited during and after the pandemic, both the staff rating and the increase in the frequency of this word in reviews suggest the development of hotel employees competences. The time of the pandemic was a period of limited ability of hotels to operate normally, which in many cases resulted in lay-offs (mainly of the line workers). Therefore, the remaining employees had to take over the duties of the dismissed staff. Subsequently, many of hotel managers tried to look for opportunities provide satisfaction services by engaging employees at all levels. In addition, the limited number of employees required an introduction of model of multi-functionality and multi-tasking. As a result, employees had the opportunity to develop their competences in many areas related to hotel management – not only during the crisis but also during recovery periods. Therefore, it can be deduced that the pandemic had a positive impact on the implementation of HR processes in hotel properties.

Subsequently, the biggest increase of score being observed in hotels with low star categories implies they were the most efficient at ensuring the quality required by guests despite facing the COVID-related difficulties and the decrease in price being mentioned in reviews might suggest a shifted priorities and value perceptions in tourists using hotels in Poland. Given the economic repercussions of the pandemic, many individuals faced financial constraints which in turn made them be more forgiving or have reduced expectations towards services offered. Practicality, safety and basic comfort might have taken precedence over luxury, leading guests to appreciate basic services more and not overly focus on frills. It could be seen especially in the increase of word *cleanness* being used in reviews during the pandemic – a finding in line with the works of Hu et al. (2021) and Sekar and Santhanam (2022). At the same time, with the increase in room score, our findings would suggest that this increase was noticed by hotel managers and adequately incorporated in their service standards.

While Hu et al. (2021) showed the decrease in importance of location/surroundings and breakfast in pandemic times, our findings show more light on this matter. With the score of hotel location increasing over the pandemic as well as its frequency increasing in the reviews, it can be deduced that customers' needs in this aspect were changed by the COVID pandemic. Such an implication would be strengthened by the findings of Xu et al. (2022), who also observed that there was no decrease in the importance of location in hotel evaluation, but rather a shift in what ways customers appreciate the surroundings of hotel properties. Our findings additionally showcased that there was no significant decline in the importance of food services during the pandemic. While in the period before the pandemic, 'breakfast' was not one of the most frequently mentioned words in guest reviews, its frequency increased significantly during and after the pandemic. It can be deduced that this change can be attributed to the development of room service in hotels that did not offer such services before the pandemic. Breakfast service delivered to the room requires greater concentration on the individual needs of guests than in the case of a buffet offer, thus its more frequent mentions in the reviews.

Conclusions

This research explored the differences in customers' behaviour by analysing changes in different reviews' attributes before, during and after COVID-19 pandemic. Through the analysis it was established that hotel attribute scores increased in all segments when comparing between pre- and during pandemic periods, as well as when comparing during and after the pandemic periods. Thus, it can be seen that despite numerous restrictions determining the scope of services offered and despite limiting the number of employees to the necessary minimum, hotels in Poland managed to maintain a high standard of services offered. This would probably not be possible without the commitment and flexibility of employees who, despite increased responsibilities, were able to cope with the changes occurring during the pandemic – also in terms of meeting guests' expectations. This process was probably mostly influenced by the continuous development of employees' competencies in the face of the challenges brought about by subsequent waves of the pandemic and the introduced restrictions.

Moreover, it was shown that lower category hotels showed the greatest flexibility, as evidenced by the highest increases in the ratings of their individual attributes in the reviews. This result is not surprising, as lower category hotels are assumed to possess a limited offer of services, rarely covering a wider range apart from the basic ones. Therefore, the enforced pandemic restrictions did not significantly affect the scope of services offered, but rather only reduced the number of rooms offered. Thus, with lower hotel occupancy, focusing on providing the highest level of possible service was not a major challenge for these hotels – even with a limited number of employees.

Our research also showed that the importance of cleanliness has increased during and in the post-pandemic period. Taking into account the high risk of infection resulting from the spread of COVID-19 pandemic, increasing the importance of this attribute of hotels during the pandemic seems natural. However, our research showcased that the importance of the cleanliness attribute for hotel guests in Poland has not decreased after the end of the pandemic, and the frequency of references to cleanliness attributes in the content of reviews remained increasing. Therefore, it can be assumed that the importance of this hotel attribute for guests will remain high for a long time. This in turn would oblige hotels to maintain their attention on meeting hygiene and sanitary standards at the highest possible level.

Contributions for research

By the given study, it was possible to improve the way one can look at and analyse online reviews, blending both quantitative studies and content analysis of electronic word-of-mouth and achieving a richer understanding of hotel guests' sentiment. A significant finding from this study highlights the value of Google Reviews for hotel-related research. When compared to sites like TripAdvisor and Booking.com, Google Reviews offers a broader and more varied perspective for managerial studies in many different fields. Another contribution of this research is its focus on the progression of online reviews over time: from before the pandemic, through its peak and into the post-pandemic phase. Our research stands out in the literature as one of the pioneering studies which investigated eWOM originating before, during and after the COVID-19. By analysing such an extensive dataset, it was possible to expose the shifting sentiments of the public during these challenging times, provide businesses with valuable insights and cover a significant research gap.

Contributions for practice

This research revealed the topics and aspects that are currently holding the highest importance for customers, guiding hotels to prioritise the most critical aspects of a guest's stay. At the same time, the observed increase in general scores indicates that the measures taken by hotels during the pandemic reso-nated well with guests, suggesting that similar actions would be beneficial in the future should a similar crisis situation arise. Furthermore, the differentia-tion in review score trends across all hotel categories provides valuable insights for hotel managers by allowing them to tailor their strategies and adapt their properties based on their hotel's star rating.

Limitations

There are some limitations to this research. First, this investigation only included reviews of hotels that survived the pandemic, which may slightly dis-tort the overall picture of the results. In future research, it is worth analysing the opinions of hotels existing before the pandemic, but also those newly estab-lished during and after the COVID-19 pandemic. Secondly, the given research did not compare between the changes in negative and positive eWOM. In the future, reviews should be analysed in this respect to verify and supplement the conclusions of this study. The given study's ambiguous results concerning hotels' location evaluation call for further research in this area as it may expand the current perspective on the importance of location in making investment decisions.

References

AHLA. (2020). *Leisure and Hospitality Industry Proves Hardest Hit by COVID-19.* https://www.ahla.com/covid-19s-impact-hotel-industry

Anagnostopoulou, S. C., Buhalis, D., Kountouri, I., & Manousakis, E. (2020). The impact of online reputation on hotel profitability. *International Journal of Contemporary Hospitality Management*, January. https://doi.org/10.1108/IJCHM-03-2019-0247

Baum, T., Cheung, C., Kong, H., Kralj, A., Mooney, S., Thi Thanh, H. N., Ramachandran, S., Ružic, M. D., & Siow, M. L. (2016). Sustainability and the tourism and hospitality workforce: A thematic analysis. *Sustainability (Switzerland)*, 8(8). https://doi.org/10.3390/su8080809

Baum, T., Mooney, S. K. K., Robinson, R. N. S., & Solnet, D. (2020). COVID-19's impact on the hospitality workforce – New crisis or amplification of the norm? *International Journal of Contemporary Hospitality Management*, 32(9), 2813–2829. https://doi.org/10.1108/IJCHM-04-2020-0314

Bhaiswar, R. (2020). Evolution of electronic word of mouth: A systematic literature review using bibliometric analysis of 20 years. *FIIB Business Review*, 10(3), 215–231. https://doi.org/10.1177/23197145211032408

Casado-Díaz, J. M., & Simón, H. (2016). Wage differences in the hospitality sector. *Tourism Management*, 52, 96–109. https://doi.org/10.1016/j.tourman.2015.06.015

Cushman & Wakefield. (2023). *Hospitality Operator Beat 2023.* https://www.cushmanwakefield.com/en/poland/news/2023/09/hospitality-operator-beat-survey-h1-2023

Dombey, O. (2004). The effects of SARS on the Chinese tourism industry. *Journal of Vacation Marketing*, 10(1), 4–10. https://doi.org/10.1177/135676670301000101

Filieri, R. (2016). What makes an online consumer review trustworthy? *Annals of Tourism Research, 58*(3), 46–64. https://doi.org/10.1016/j.annals.2015.12.019

Gerdt, S. O., Wagner, E., & Schewe, G. (2019). The relationship between sustainability and customer satisfaction in hospitality: An explorative investigation using eWOM as a data source. *Tourism Management, 74*(December 2018), 155–172. https://doi.org/10.1016/j.tourman.2019.02.010

Gretzel, U., Fuchs, M., Baggio, R., Hoepken, W., Law, R., Neidhardt, J., Pesonen, J., Zanker, M., & Xiang, Z. (2020). e-Tourism beyond COVID-19: A call for transformative research. *Information Technology and Tourism, 22*(2), 187–203. https://doi.org/10.1007/s40558-020-00181-3

Henderson, J. C., & Ng, A. (2004). Responding to crisis: severe acute respiratory syndrome (SARS) and hotels in Singapore. *International Journal of Tourism Research, 6*(6), 411–419. https://doi.org/10.1002/jtr.505

Hospitality Europe. (2017). *The hospitality industry's contributions to European economy society.* www.hotrec.eu/facts-figures-2

HOTREC. (2020). *HOTREC Annual Report 2020-2021.* https://www.hotrec.eu/hotrec-annual-report-2020-2021/

Hu, F., Teichert, T., Deng, S., Liu, Y., & Zhou, G. (2021). Dealing with pandemics: An investigation of the effects of COVID-19 on customers' evaluations of hospitality services. *Tourism Management, 85*(March), 104320. https://doi.org/10.1016/j.tourman.2021.104320

Huete-Alcocer, N. (2017). A literature review of word of mouth and electronic word of mouth: Implications for consumer behavior. *Frontiers in Physiology, 8*(JUL), 1–4. https://doi.org/10.3389/fpsyg.2017.01256

Jang, S., Kim, J., Kim, J., & Kim, S. (2021). COVID-19 and peer-to-peer accommodation: A spatial and experimental approach to Airbnb consumption. *Journal of Destination Marketing and Management, 20*, 100563.

Jezierski, A. (2022). Determinanty satysfakcji z usług polskich hoteli – analiza ocen z Booking.com. *Studia Periegetica, 2*(38), 9–30. https://doi.org/10.5604/01.3001.0015.9188

Jezierski, A., Wszendybył-Skulska, E., & Kopera, S. (2022). Crisis-resistant tourists – A study of hotel online reviews in the times of COVID-19. *Polish Journal of Sport and Tourism, 29*(4), 29–36. https://doi.org/10.2478/pjst-2022-0024

Kim, S. S., Chun, H., & Lee, H. (2005). The effects of SARS on the Korean hotel industry and measures to overcome the crisis: A case study of six Korean five-star hotels. *Asia Pacific Journal of Tourism Research, 10*(4). https://doi.org/10.1080/10941660500363694

Kim, W. G., Lim, H., & Brymer, R. A. (2015). The effectiveness of managing social media on hotel performance. *International Journal of Hospitality Management, 44*, 165–171. https://doi.org/10.1016/j.ijhm.2014.10.014

Koronawirus: informacje i zalecenia. (2023). https://www.gov.pl/web/koronawirus/aktualne-zasady-i-ograniczenia

Kosová, R., & Enz, C. A. (2012). The Terrorist Attacks of 9/11 and the Financial Crisis of 2008. *Cornell Hospitality Quarterly, 53*(4), 308–325. https://doi.org/10.1177/1938965512457021

Ladhari, R., & Michaud, M. (2015). EWOM effects on hotel booking intentions, attitudes, trust, and website perceptions. *International Journal of Hospitality Management, 46*, 36–45. https://doi.org/10.1016/j.ijhm.2015.01.010

Łapko, A., Panasiuk, A., & Strulak-Wójcikiewicz, R. (2020). The state of air pollution as a factor determining the assessment of a city's tourist attractiveness — Based on the opinions of Polish respondents. *Sustainability, 12*(1466). https://doi.org/10.3390/su12041466

Lee, H., Min, J., & Yuan, J. (2021). The influence of eWOM on intentions for booking luxury hotels by Generation Y. *Journal of Vacation Marketing, 27*(3), 237–251. https://doi.org/10.1177/1356766720987872

Liu, K.-N., & Hu, C. (2022). Critical success factors of green hotel investment in Taiwan. *International Journal of Contemporary Hospitality Management, 34*(3), 951–971. https://doi.org/10.1108/IJCHM-03-2021-0368

Maphanga, P. M., & Henama, U. S. (2019). The tourism impact of Ebola in Africa: Lessons on crisis management. *African Journal of Hospitality Tourism and Leisure, 8*(3), 1–13.

Marchante, A., Ortega, B., & Pagan, R. (2005). Educational mismatch and wages in the hospitality sector. *Tourism Economics, 11*(1), 103–117.

Mathews, S., Prentice, C., Tsou, A., Weeks, C., Tam, L., & Luck, E. (2021). Managing eWOM for hotel performance. *Journal of Global Scholars of Marketing Science, 32*(3), 1–20. https://doi.org/10.1080/21639159.2020.1808844

Mehta, M. P., Kumar, G., & Ramkumar, M. (2021). Customer expectations in the hotel industry during the COVID-19 pandemic: A global perspective using sentiment analysis. *Tourism Recreation Research*, 1–18. https://doi.org/10.1080/02508281.2021.1894692

Mishra, A., & Satish, S. M. (2016). eWOM: Extant research review and future research avenues. *VIKALPA - The Journal for Decision Makers, 41*(3), 222–233. https://doi.org/10.1177/0256090916650952

Mizrachi, I., & Fuchs, G. (2016). Journal of hospitality and tourism management should we cancel? An examination of risk handling in travel social media before visiting ebola-free destinations. *Journal of Hospitality and Tourism Management*, 1–7. https://doi.org/10.1016/j.jhtm.2016.01.009

Naab, T. K., & Sehl, A. (2017). Studies of user-generated content: A systematic review. *Journalism, 18*(10), 1256–1273. https://doi.org/10.1177/1464884916673557

Nilashi, M., Ali Abumalloh, R., Alrizq, M., Alghamdi, A., Samad, S., Almulihi, A., Althobaiti, M. M., Yousoof Ismail, M., & Mohd, S. (2022). What is the impact of eWOM in social network sites on travel decision-making during the COVID-19 outbreak? A two-stage methodology. *Telematics and Informatics, 69*(August 2021), 101795. https://doi.org/10.1016/j.tele.2022.101795

Nwobi, F. N., & Akanno, F. C. (2021). Power comparison of ANOVA and Kruskal–Wallis tests when error assumptions are violated. *Metodoloski Zvezki, 18*(2), 53–71. https://doi.org/10.51936/LTGT2135

Olivieri, M., Colleoni, E., & Bonaccorso, G. (2023). How have travelers' needs evolved because of the COVID-19 pandemic? Corporate reputation building in tourism industry on digital platforms. In R. Rialti, Z. Kvítková, & T. Makovník (Eds.), *Online reputation management in destination and hospitality* (pp. 51–71). Emerald Publishing Limited. https://doi.org/10.1108/978-1-80382-375-120231003

Panasiuk, A. (2013). Marka turystyczna jako instrument zarządzania regionalną gospodarką turystyczną w warunkach sytuacji kryzysowych. *Wspolczesne Zarzadzanie Kwartalnik Srodowisk Naukowych i Liderów Biznesu, 12*(1), 21–30. https://doi.org/10.5604/16435494.1052269

Pine, R., & Mckercher, B. (2004). Research in brief The impact of SARS on Hong Kong's tourism industry. *International Journal of Contemporary Hospitality Management*, *16*(2), 139–143. https://doi.org/10.1108/09596110410520034

Podhorodecka, K., & Bąk-Filipek, E. (2022). Zmiany na rynku pracy w sektorze turystycznym w Polsce w dobie pandemii covid-19. *Rynek Pracy*, *182*(3), 6–18. https://doi.org/10.5604/01.3001.0016.0334

Ritchie, B. W. (2004). Chaos, crises and disasters: A strategic approach to crisis management in the tourism industry. *Tourism Management*, *25*(6), 669–683. https://doi.org/10.1016/j.tourman.2003.09.004

Robinson, R., Brenner, M., & Temesgen, K. (2020). "Canary in the coalmine": Hospitality worker satisfaction and wellbeing compared to all industries. In *CAUTHE 2020: 20: 20 Vision: New Perspectives on the Diversity of Hospitality, Tourism and Events*. Auckland University of Technology.

Rodríguez, A. (2002). Determining factors in entry choice for international expansion. The case of the Spanish hotel industry. *Tourism Management*, *23*(6), 597–607. https://doi.org/10.1016/S0261-5177(02)00024-9

Rosario, A. B., de Valck, K., & Sotgiu, F. (2020). Conceptualizing the electronic word-of-mouth process: What we know and need to know about eWOM creation, exposure, and evaluation. *Journal of the Academy of Marketing Science*, *48*, 422–448. https://doi.org/10.1007/s11747-019-00706-1

Sekar, S., & Santhanam, N. (2022). Effect of COVID-19: Understanding customer's evaluation on hotel and airline sector—A text mining approach. *Global Business Review*. https://doi.org/10.1177/09721509221106836

Serra-Cantallops, A., Ramon-Cardona, J., & Salvi, F. (2018). The impact of positive emotional experiences on eWOM generation and loyalty. *Spanish Journal of Marketing - ESIC*, *22*(2), 142–162. https://doi.org/10.1108/SJME-03-2018-0009

Shin, H., & Kang, J. (2020). Reducing perceived health risk to attract hotel customers in the COVID-19 pandemic era: Focused on technology innovation for social distancing and cleanliness. *International Journal of Hospitality Management*, *91*(June), 102664. https://doi.org/10.1016/j.ijhm.2020.102664

SimilarWeb - Website analytics. (2023). https://pro.similarweb.com

Statista. (2020). *Market size of the hotel and resort industry worldwide from 2011 to 2020, with a forecast for 2021*. Market size of the hotel and resort industry worldwide from 2011 to 2020, with a forecast for 2021.

STR. (2022). *Smith Travel Reports*. www.str.com

Tapia, M., & Alberti, G. (2019). Unpacking the category of migrant workers in trade union research: A multi-level approach to migrant intersectionalities. *Work, Employment and Society*, *33*(2). https://doi.org/10.1177/0950017018780589

Williamson, D., & Harris, C. (2019). Talent management and unions: The impact of the New Zealand hotel workers union on talent management in hotels (1950–1995). *International Journal of Contemporary Hospitality Management*, *31*(10), 3838–3854. https://doi.org/10.1108/IJCHM-10-2018-0877

Wszendybył-Skulska, E. (2012). *Human Capital in Hotel Industry*. CeDeWu.

Wszendybył-Skulska, E. (2013). Inwestycje w kapitał ludzki - sposobem na kryzys przedsiębiorstw hotelarskich. In M. Bednarczyk & E. Wszendybył-Skulska (Eds.), *Zarządzanie turystyką w kryzysie: edukacja i marka*. CeDeWu.

Xie, K., Kwok, L., & Wang, W. (2017). Monetizing managerial responses on TripAdvisor: Performance implications across hotel classes. *Cornell Hospitality Quarterly*, *58*(3), 240–252. https://doi.org/10.1177/1938965516686109

Xu, J., Wang, X., Zhang, J., Huang, S. (Sam), & Lu, X. (2022). Explaining customer satisfaction via hotel reviews: A comparison between pre- and post-COVID-19 reviews. *Journal of Hospitality and Tourism Management, 53*(July), 208–213. https://doi.org/10.1016/j.jhtm.2022.11.003

Xu, X., & Li, Y. (2016). The antecedents of customer satisfaction and dissatisfaction toward various types of hotels: A text mining approach. *International Journal of Hospitality Management, 55*, 57–69. https://doi.org/10.1016/j.ijhm.2016.03.003

Yen, C. L. A., & Tang, C. H. H. (2019). The effects of hotel attribute performance on electronic word-of-mouth (eWOM) behaviors. *International Journal of Hospitality Management, 76*(September 2017), 9–18. https://doi.org/10.1016/j.ijhm.2018.03.006

Yu, M., Cheng, M., Yang, L., & Yu, Z. (2022). Hotel guest satisfaction during COVID-19 outbreak: The moderating role of crisis response strategy. *Tourism Management, 93*(June), 104618. https://doi.org/10.1016/j.tourman.2022.104618

Zaman, M., Botti, L., & Thanh, T. V. (2016). Weight of criteria in hotel selection: An empirical illustration based on TripAdvisor criteria. *European Journal of Tourism Research, 13*, 132–138.

Zervas, G., Proserpio, D., & Byers, J. W. (2021). A first look at online reputation on Airbnb, where every stay is above average. *Marketing Letters, 32*(1), 1–16. https://doi.org/10.1007/s11002-020-09546-4

3

CROWDING'S IMPACT ON TOURIST EXPERIENCE IN DIVERSE DESTINATIONS

A comparative study of Las Vegas and the Great Smoky Mountains

Nicole Hankins, Akshaya Pawar and Anita Zatori

Introduction

The possible impacts of overcrowding are diversified when it comes to tourism. Crowding and more specifically overcrowding might have many negative effects, including environmental degradation, directly and indirectly decreasing the quality-of-life local communities, but also a negative impacting the tourist experience leading to reduced satisfaction with the destination or its revisit intention.

Overconsumption of natural resources, accumulation of waste and pollution, and damage to the environment are some of the major impacts of overcrowding. Overcrowding may negatively impact the local communities as it can lead to a burden on infrastructure and a degradation of the quality of life for residents. It can be the cause of service quality degradation, as tourism providers struggle to meet the demands of large numbers of tourists; furthermore, overcrowding can make it difficult for tourists to access popular tourist attractions, leading to frustration and disappointment (McKinsey & Company, 2017). Consequently, overcrowding can also lead to reduced satisfaction with the tourist experience, as tourists may feel that they are not able to fully enjoy the destination due to the crowds (Ramkissoon et al., 2022).

The consequences of overtourism on host communities are well articulated in the literature. For example, Dodds and Butler (2019) discuss the diminishing sense of place among residents, while Milano, Novelli and Cheer (2019) point out the issue of permanent changes to lifestyles. While the environmental and sociocultural impacts of overcrowding have been extensively studied, there is much less academic discussion on how overcrowding is affecting the visitor experience itself. The degradation of tourist experience quality due to overtourism

DOI: 10.4324/9781032637778-4

was brought to our attention in some recent studies (e.g., Capocchi et al. 2019; Koens, Postma, and Papp 2018). However, a small number of studies place the impact of overcrowding on visitor experience in their focal point of research. These studies (Papadopoulou et al., 2023; Tokarchuk et al., 2022; Yu and Egger, 2020) impartially call for further research for the area and theoretical and managerial discourse. The study aims to address this research gap.

Crowding and overcrowding have become an issue for residents as well as tourists in destinations as diverse as Barcelona, Amsterdam, New Orleans, Dubrovnik or the island of Koh Phi Phi in Thailand. Although overcrowding can be an issue in destinations of diverse characters, the dynamics of overcrowding (associated with the impact of crowding on host communities and the tourists themselves) in diverse destination character settings remain unclear.

The study in this chapter addresses this issue and aims to explore how overcrowding influences tourist experiences in destinations with distinct characters. The focus is on two distinct destinations: Las Vegas, known as a tourism hotspot and epicentre, and the tourist destination of the Great Smoky Mountains, a popular destination with a multi-centre character. The two destinations are two of out of ten US tourist destinations on the 'Map of Overtourism' by Responsible Travel (Francis, 2023).

Literature review

'Crowding' and 'overcrowding' are terms used in tourism discourse to describe a situation resulting from a disproportionately large influx of visitors within an area (Goodwin, 2017). These two terminologies are both used to describe the above phenomenon, and while they are slightly different there is no agreement or definition on what level of crowding becomes overcrowding. Thus, they are both used in the recent study interchangeably.

Whether tourism should follow a growth model has been questioned in the last decade as a consequence of continued rapid growth in tourist arrivals globally, but especially in popular destinations. The impactful and instant power of social media has intensified crowding problems at popularised tourist attractions and destinations.

The negative impacts of tourism overcrowding have started to overshadow the other benefits of tourism as an industry and intercultural phenomenon at attraction sites with intensified influx of visitors. At intensively visited tourist attractions a more critical perception of crowdedness emerges that does not only impact the quality of life of the residents but has a potential to deteriorate the quality of the tourist experience of visitors, too. What is the degree of crowdedness to be perceived as overcrowding with a negative impact on the tourist experience? What is the 'moment of truth' that turns an exciting destination vibe into an overcrowding experience?

Crowding

Concerns about crowding at destinations, particularly tourist sites, have been a recurring theme in tourism research for over a half century. Crowding can be defined as a psychological evaluation of the perceived human density in a specific area (Shelby, Vaske, and Heberlein 1989; Stokols 1972).

The concept of the crowding effect was initially formulated to describe the emotions that result from the perceived density of people and the resulting limitations (Stokols, 1972). Turner and Ash (1975) referred to 'golden hordes', and this concept was popularised through Doxey's (1975) 'Irritation Index'. Subsequently, Social Exchange Theory (Emerson, 1976) pointed out how tourism crowding can lead to a conflict between hosts and guests.

In tourism theory, there is a consensus that crowding is situation-dependent, influenced by individual norms, values and the perceived characteristics of fellow tourists. Tourists tend to associate crowding with waiting, stress or negative qualities of a destination (Perdue et al., 1999).

It is worth noting that some researchers have highlighted the potential for crowding to have positive aspects (Choi, Mirjafari, and Weaver, 1976), although this perspective has often been overlooked (Neuts and Nijkamp, 2012; Oklevik et al., 2019). Notions of being in the right place, particularly in the broader context of following trends, have gained significance due to information cascades, the prevalence of news on social media (Turkle, 2015).

Overtourism

While the term is relatively recent, many of the problems involved have a long history, particularly in well-visited destinations. Dodds and Butler (2019) point out that overtourism is a new term for an old problem, namely, excessive numbers of tourists at a specific destination that can result in negative impacts of all types on the community involved.

The concept of carrying capacity introduces the idea of overtourism, which highlights that 'the volume of tourists and their conduct can overburden the locations they explore, causing harm to both the tourism assets and the daily lives of the residents in destination regions' (Wall, 2020, p. 212).

Goodwin (2011) defines overtourism as a scenario where either residents or tourists perceive a place as excessively visited, resulting in a transformation of its character, a loss of its authenticity (primarily for visitors) and causing annoyance and irritation (for residents). According to Koens et al. (2018), overtourism is characterised as an excessive adverse impact of tourism on the host communities and/or the natural environment.

Despite the urgency and the fact that much literature has been produced, overtourism is still not fully defined academically and remains subject to various interpretations (Koens et al., 2018; Dodds and Butler, 2019; Mihalic, 2020).

Butler (2018) makes a clear distinction between the concept of overtourism and that of overcrowding. He clarifies that overtourism is not synonymous with overcrowding or busy destinations, but rather signifies a situation where the number of visitors overwhelms the available services and facilities, causing significant inconvenience to permanent residents in those areas.

Koens et al. (2018) found that the typical issues caused by overcrowding include: (1) overcrowding in city's public spaces, (2) pervasiveness of visitor impact due to inappropriate behaviour, (3) physical touristifaction of city centres and other highly visited sites, (4) residents pushed out of residential areas due to Airbnb and similar platforms, and (5) pressure on local environment. However, it is worth pointing out that their research is geographically limited to European tourist hotspots.

Overcrowding's impact on tourist experience

Existing research explores the impact of overcrowding on the tourist experience (Jacobsen, Iversen, and Hem 2019; Popp 2012; Weber et al. 2017). Although residents are typically the first to voice concerns about the detrimental effects of high tourist densities on their way of life and the local environment (Gössling, McCabe, and Chen 2020), tourists have also begun to express similar concerns due to the undesirable consequences of overtourism on their own destination experience with implications for tourist satisfaction and return intent (Koens, Postma, and Papp 2018).

Excessive crowding has the potential to adversely affect both the image of a destination and the overall experience of tourists (Namberger et al. 2019). Recent research indicates that overcrowding can lead to unfavourable reactions from travellers (Jacobsen, Iversen, and Hem 2019; Machleit, Eroglu, and Mantel 2000), resulting in reduced intentions to revisit or recommend the destination (Jurado, Damian, and Fernandez-Morales 2013; Neuts and Nijkamp 2012).

Social interference theory has important implications for understanding how social cues in the tourism environment can impact cognitive performance and decision-making. In the tourism context, social cues can include interactions with other tourists, the behaviour and actions of locals, and the general atmosphere of the destination (Kwon, Ha, and Im, 2016).

Overcrowding is generally considered a negative phenomenon in tourism, as it can lead to negative consequences for both tourists and locals. However, crowding does not always entail a negative assessment of the experience, indeed, indications of positive crowding perceptions also exist in the literature (Neuts and Nijkamp 2012; Popp 2012).

Crowding may have a positive impact on the tourist experience in certain circumstances such as increased social interaction and economic benefits for the destination. In the first case, the positive impact demonstrates itself in the form of co-creation between tourists' social interaction, i.e., increased social interaction which can enhance the tourist experience quality and memorability. Tourists may have the opportunity to meet new people, share experiences and form new friendships and co-create value together this way. Popp (2012) proposed that good crowding, during a high-density visit, might occur due to human-related motivations, for example, sharing experiences with people, socialising and observing others, or undertaking group activities.

Overcrowding can also bring economic benefits to the destination, such as increased revenue, job creation and investment in infrastructure (Hootboard, 2023). These benefits can contribute to the development of the destination and enhance the quality of the tourism experience in the long run.

Crowd tolerance

The negative impact of crowding on the tourist experience is largely dependent on the perception of crowding. When the capacity in one place is exceeded, the quality of life of residents and tourist experiences might decline (Yu and Egger, 2020). However, it should be highlighted that perception of crowding is subjective and can vary depending on the individual's expectations and preferences (Kim and Yoon, 2020).

The type of tourism product the visitors participate in and their tourist motivation forms their expectations, thus it impacts their perception of crowding and crowd tolerance in situ.

Several studies reveal substantial individual differences among tourists in terms of their tolerance for crowding. For instance, Jurado et al. (2013) discovered that high-quality tourists exhibit less tolerance for crowding, while Jacobsen et al. (2019) observed that cruise passengers are more accepting of crowded conditions compared to self-organised travellers.

On the other hand, crowd tolerance as a variable's impact on crowding perception is dependent on further factors, as well, such as personality traits or the consumed tourism product (e.g., cruise).

Tourists' awareness of overcrowding in tourist hotspots can change their tourist behaviour. Papadopoulou, Ribeiro and Prayag (2023), for instance, point out that visitors might be able to develop positive coping mechanisms to crowding and become more resilient to such experiences having less of an effect on their intention to revisit or recommend the destination. But can crowd tolerance and destination character be interrelated, as well? Does the destination choice impact the level of crowd tolerance in situ?

Destination character

Most of the studies concerning crowding have found a negative relationship between the level of destination usage and destination satisfaction. Higher perceptions of crowding tend to evoke a more negative response, leading to lower levels of satisfaction (Neuts and Nijkamp, 2012; Rathnayake, 2015; Zhang et al., 2017).

Overtourism and overcrowding have been the subjects of investigation in tourist hotspots (Jacobsen, Iversen, and Hem 2019), urban areas and cities (Koens, Postma, and Papp 2018; Peters et al. 2018; Vaske and Shelby 2008). Other research on crowding has been conducted in natural parks and protected areas, where tourists seek to enjoy nature. Consequently, the ideal condition in these studies is often seen as having zero crowding, with a sharply negative norm curve (Vaske and Shelby, 2008; Rathnayake, 2015). However, norm curves observed in urban settings may display a different pattern. For example, in a study conducted by Jacobsen, Iversen and Hem (2019) in historic city centres, visitors displayed a positive assessment of crowding. Their research points out that crowd tolerance and destination character are interrelated but have not received enough attention in the academia and industry, and both elaboration on these findings and further evidence are missing. Vaske and Shelby (2008) highlighted that the perception of crowding and tolerance towards it can be highly dependent on the specific context.

Tourist hotspots in cities vs. in other destination types

Tourist hotspot destinations are typically renowned mass tourism locations that attract vacationers for short visits, and this can impact the livelihood of the host communities (Jacobsen, Iversen, and Hem 2019). The substantial influx of annual tourist arrivals not only raises concerns about sustainability but also challenges the carrying capacity of these destinations. Tourist hotspots are heavily reliant on tourism for their economic well-being and have, in some cases, reached their social and environmental carrying capacity, highlighting the phenomenon of overtourism (Insch 2020; Peters et al. 2018).

Some epicentres of tourism have been popular for years and will most likely continue to be popular for years to come. Some of these locations include New York City, Tokyo or London. Other destinations have recently become more popular tourist destinations with the rise of social media. Travellers are coming to these lesser-known locations and promoting them, whether that be through blog posts or social media. Some examples of these locations include Alberta, Canada; Reykjavik, Iceland; Chefchaouen, Morocco; and Cambria, California (Cranley, 2019).

Several smaller towns with rich and unique heritage in Europe experience problems due to overcrowding, affecting both the quality-of-life of residents and the tourist experience. Such destinations are, for example, Hallstat in Austria (Vlamis, 2023), Dubrovnik in Croatia or Portofino in Italy (TasteAtlas, 2023).

In the United States, some national parks experience issues due to overcrowding. A recent report from 2018 describes the visitor experience in certain parks as 'insanely crowded, almost like a theme park' (Peltier/Skift, 2018). The report highlights how some parks, such as Zion or Arches, are unable to deliver the natural experience the park service wants visitors to have, and in some parks, shuttle buses are beyond capacity and are dumping huge numbers of people onto trails all at once, while parking has been an issue in several parks, including the Great Smoky Mountains.

Las Vegas

Las Vegas, Nevada, is one of the most travelled places in the United States and has been a tourism hotspot since the 1970s. In 1970, Las Vegas received around 6.7 million tourists. In 2022, Las Vegas received 38.8 million tourists (falling 3.7 million visitors short of the 2019 visitation total) (LVCVA, 2023).

In the early 1990s, Spanier (1992) characterised Las Vegas as a city pulsating with vitality, while Douglass and Raento (2004) deemed it the most extraordinary among American urban locales. Despite the edgy image cultivated by the campaign slogan 'What happens in Vegas stays in Vegas' around the turn of the millennium, as highlighted by Wood (2005), tourism continued to flourish.

Tourism takes the top spot as the leading industry in the city. It generated nearly US$80 billion in tourism revenue in 2022 (LVCVA, 2023), and about 44% of local residents are employed in the hospitality industry (Nelson, 2018). Despite the heavy reliance on tourism for the local economy, there is a notable backlash from residents. In fact, the anti-tourism sentiment is so widespread that Las Vegas ranks among the top five 'tourism-hating' cities in the United States, according to CityLab, a ranking based on geo-tagged tweets (Gan, 2015). Interestingly, the term 'overtourism' is not a topic of discussion among Las Vegas tourism policymakers. International tourist arrivals steeply declined because of the COVID-19 pandemic, and this resulted in the loss of an estimated US$32 billion in revenue and 40,000 jobs within the national tourism industry, and Las Vegas was particularly hard-hit.

Tourism has almost completely recovered since the COVID-19 pandemic. According to the Las Vegas Convention and Visitors Authority (LVCVA, 2023), the visitor volume for March 2023 reached 40.4 million. This figure is only 5.6% below the pre-pandemic peak recorded in February 2020. Additionally, hotel occupancy rates on the Las Vegas Strip reached an impressive 90.5% in March 2023 (LVCVA, 2023). The tourism rebound is driven by special events, a mixture of high-profile sporting events and concerts.

Las Vegas is not just a vacation destination; it is also one of the top US cities for business tourism, hosting over 21,000 conventions each year in expansive venues such as the 3.2-million-square-foot Las Vegas Convention Center (Nelson, 2018). Travellers who are seeking a more traditional Vegas trip may want to go to one of their 60 major casinos (Salem, 2022). Like many other famous tourism epicentres, Las Vegas seems as though it offers something for everyone.

Some of the most famous tourist attractions include the 'Welcome to Fabulous Las Vegas' sign, now complemented by a compact parking facility to accommodate the numerous cars and buses utilised by visitors. Predominantly, these visitors utilise the site as a backdrop for photographic purposes (Weaver,

2011). Another iconic tourist attraction is the Strip, characterised by a constant flow of pedestrian traffic, is undoubtfully a number one site to visit in Vegas, and a place that is hard to miss when visiting the city.

Great Smoky Mountains

The most popular gateway community of the Great Smoky Mountains National Park is the area of Gatlinburg and Pigeon Forge in Tennessee, United States, also known as Sevier County, Tennessee. Gatlinburg is a nationwide tourist city, and it is one of the most historic tourism venues in the United States (Lee et al., 2020). The Gatlinburg/Pigeon Forge destination is one of the most visited areas in the United States (Fishman, 2015). It stands out as a destination for both nature and culture lovers. Visitors tend to gravitate towards this location when they are in search of quality time, nature, exploration of Southern culture (especially Tennessee culture), as well as when they show any interest in the more traditional tourist attractions; such as theme parks, shopping centres, zoos, outdoor adventures, nature trails, museums and dinner shows.

Tourism growth has been very conspicuous and rapid since the early 1970s, resulting in tourism-dominated local economies. The destination benefits from being on the western side of the park, which is served by Interstate 75 serving tourists from such states as Ohio and Florida. It is within a day's drive for more than half the population of country (Lee et al., 2020). Thus, many people are likely to visit the destination due to its relative proximity.

During the 1980s Pigeon Forge witnessed the construction of outlet malls and the development of the Dollywood theme park. In the 1990s the rapid growth in Pigeon Forge continued with the introduction of country-music theatres (Tooman, 1997).

The expansion of tourism industry near the national park has also been enabled by geographic features; the Tennessee side of the mountains has allowed ample room for expansion after Gatlinburg became overcrowded and tourism spilled out first into Pigeon Forge and then into Sevierville.

The Great Smoky Mountains National Park had more than 11.3 million recreational visitors in 2016. This figure does not include the approximately 11 million travellers on the Gatlinburg–Pigeon Forge Spur. The park is the most visited national park in the United States, while the second is Grand Canyon with 4.6 million visits (National Park Service, 2023) significantly less than 11.3 million.

An important tourist attraction in this area is Dollywood, Dolly Parton's amusement park, located in Pigeon Forge, Tennessee. The theme park averages around three million visitors each year (Warrick, 2022) making it Tennessee's most popular tourist attraction. The destination offers a wide variety of popular tourist attractions, including Ripley's Aquarium of the Smokies, Ripley's Haunted Adventure, Anakeesta, Gatlinburg Sky Lift Park, Gatlinburg Space Needle, winter activities on the Ober Mountain and another six smaller amusement parks.

Furthermore, the destination is the home of about eighteen distilleries and wineries, and four outlet malls offering endless shopping opportunities.

In 2015, TripAdvisor contributors named Gatlinburg the first destination on the rise in the United States and the fourth destination on the rise in the world (Fishman, 2015). TripAdvisor also ranked Gatlinburg among the top five most Beautiful Little Mountain Towns in America and pinpointed that Gatlinburg is a nature-friendly tourist city for the Great Smoky Mountains National Park, which represents a living tourism history of the United States (TripAdvisor.com, 2016).

A study by Lee, Benjamin and Childs (2020) classified tourists' emotional responses and sentiments manifested in the TripAdvisor reviews of Gatlinburg tourist attractions. The results of this study revealed that travellers were more likely to have favourable emotional experiences towards the attractions. Their findings imply that Gatlinburg, a nationwide tourist city, appears to likely attract and retain visitors.

Methodology

Data

The study employed a self-administered questionnaire as a main method of enquiry. The questionnaires were distributed through Qualtrics and Prolific in two phases: from March to April 2023 and from August to October 2023. Participants were screened using screening questions to ensure that only participants who had visited the two destinations (Great Smoky Mountains and Las Vegas) could participate. A total number of 166 valid questionnaires were used for consecutive analysis. Considering the focus of the study, the survey instrument consisted of three parts. The first section assessed the overcrowding and experience. The second part assessed the purpose and intentions of the visit. Finally, the third part in the survey instrument included the assessment of social-demographic information such as gender, age, travel partner and nationality.

The survey instrument measured 'Overcrowding' as the number of places where respondents experienced crowds via the variable 'Number of crowded places'. The type of impact on tourist experience was measured on a five-point Likert scale from 'Definitely No' to 'Definitely Yes' (via the survey item 'Do you think that your tourist experience would have been better if there were less people in your previously specified overcrowded areas?'). The overcrowding's impact on tourist experience was also measured as 'Impact on Tourist Experience' on a five-point Likert scale from 'Only negative' to 'Only positive' (via the survey item 'Did the overcrowding that you experienced have any effect on your tourist experience?'). During the final coding, the 'Impact on Tourist Experience' was recoded as a nominal variable with three categories – negative impact, no impact and positive impact – to ascertain the direction of the impact of overcrowding. The variables are recorded in Table 3.1.

TABLE 3.1 Variable list

Construct	Variable name	Survey item	Measurement	
Overcrowding	**Number of crowded places**	If you answered yes or maybe to the previous question, please specify the areas in which you noticed that overcrowding was greater:	Specified number of places	
Type of Impact on experience	**Type of Impact on experience**	Did the overcrowding that you experienced have any effect on your tourist experience?	1	It only had a negative effect
			2	It had a rather negative effect
			3	Undecided
			4	It had a rather positive effect
			5	It only had a positive effect
			Recoded as:	
			1	Negative
			2	Undecided
			3	Positive
Crowd Tolerance	**Crowd Tolerance**	In general, do you mind being in crowded places?	1	Do not mind at all
			2	Somewhat mind
			3	Find them extremely difficult to be in
Tourist experience	**Impact on tourist experience**	Do you think that your tourist experience would have been better if there were less people in your previously specified overcrowded areas?	1	Definitely not (low impact)
			2	Probably not
			3	Might or might not
			4	Probably yes
			5	Definitely yes (high impact)
Revisit Intention	**Revisit Intention**	Would you consider revisiting the same location if it were just as crowded as your last visit?	1	Not at all
			2	Probably not
			3	Might or might not
			4	Probably yes
			5	Absolutely
Trip Satisfaction	**Trip Satisfaction**	Overall, how satisfied were you with your trip?	1	Extremely dissatisfied
			2	Somewhat dissatisfied
			3	Neither satisfied nor dissatisfied
			4	Somewhat satisfied
			5	Extremely satisfied
Trip Memorability	**Trip Memorability**	Overall, was your trip to this destination memorable in a positive or a negative way?	1	Extremely negative
			2	Somewhat negative
			3	Neither positive nor negative
			4	Somewhat positive
			5	Extremely positive

IBM SPSS 29.0 and Excel were employed to describe the data and to test the models. Linear regression approach was applied while the moderating effects of destination character, crowd tolerance and impact on experience were tested using Model 1 with PROCESS macro. According to Hayes's suggestion (2013), the bootstrapping method with 5,000 re-samples and a bias-corrected 95% confidence interval was used to examine the significance of indirect effects.

Results

Common method bias assessment

As the study is cross-sectional in nature, and the main variables were collected as part of one survey from the same group of respondents, the results may be susceptible to common method bias (CMB) (Kock et al., 2021; Podsakoff et al., 2012). To mitigate the effects of CMB, some procedural controls were employed. Procedural controls focused on a rigorous survey design, including providing clear instructions to participants, ensuring anonymity of responses and keeping the question items simple and concise (Kock et al., 2021).

Respondents profile

The findings showed that the majority of the respondents fell between the ages of 18 and 58 (92.8%) while only 7.2% of respondents were older than 58 and older. Over half of the respondents in the sample were female (58.4%). When compared between the two destinations, Great Smoky Mountains had higher percentage of female respondents (68.75%) vis-à-vis Vegas which had almost even male (47%) and female (53%) participants. On average more respondents travelled with their family (33.1%), with friends (25.3%) or with their spouse (20.5%). Not surprisingly, given the destinations profile, a higher number of Great Smoky Mountains participants travelled to the destination with spouse and kids (41%) whereas more Vegas respondents had friends as travel partners (35%) during their trip.

Further, the study also collected data on the destination interactions (type of overcrowding experience) of the visitors on a four-point scale (1-Not at all 2-Somewhat 3-Often 4-A lot). When compared between Great Smoky Mountains and Las Vegas, Great Smoky Mountains visitors consistently reported more negative destination interactions caused by crowding than Las Vega visitors. On average, Great Smoky Mountains visitors reported a higher incidence of long wait times (M = 2.78 compared to 2.16 in Vegas), not being able to visit the attraction properly (M = 2.14 compared to 1.83 in Vegas) and being unable to 'enjoy the moments' (M = 2.21 compared to 1.85 in Vegas) than

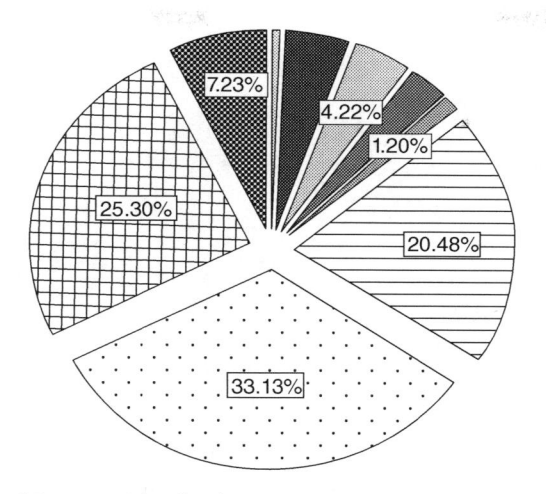

FIGURE 3.1 Travel Partner Distribution.

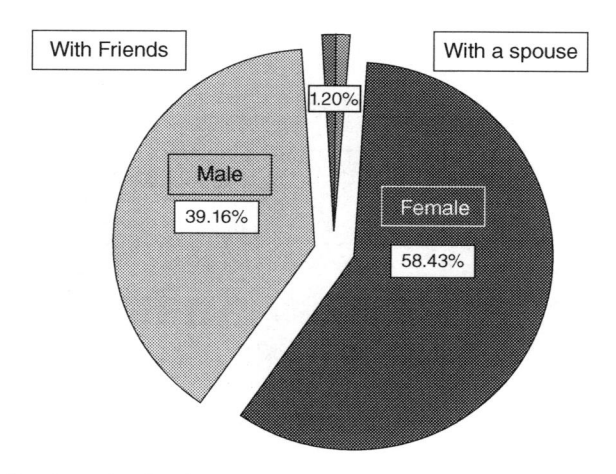

FIGURE 3.2 Gender Distribution.

tourists in Las Vegas. Overall, Great Smoky Mountains had higher aggregate scores on negative impacts of overcrowding on tourist experience (aggregate M = 2.45) than Las Vegas (aggregate M = 1.99), while 'Invaded personal space' (aggregate M = 2.48) and 'Long wait times' (aggregate M = 2.47) were the most significant issues. The findings are presented in Table 3.2.

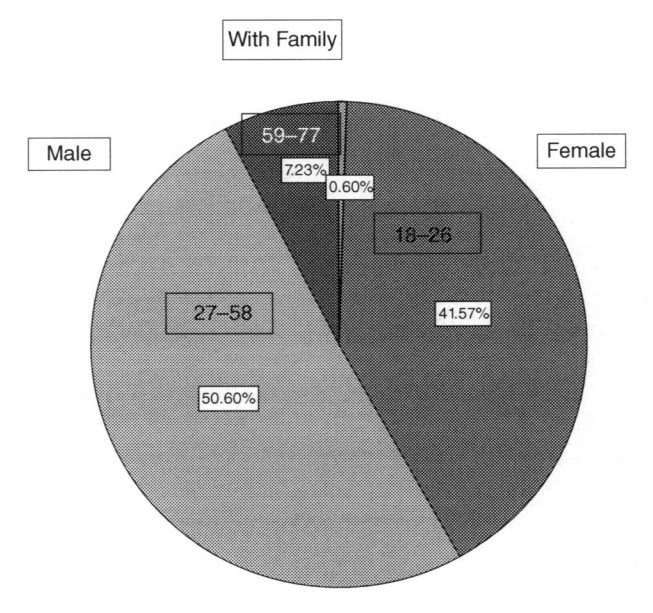

FIGURE 3.3 Age Distribution.

TABLE 3.2 Descriptive statistics

Experience	Great Smoky Mountains N = 80				Las Vegas N = 85			
	Min	Max	Mean	Std. Deviation	Min	Max	Mean	Std. Deviation
Long wait times	1	4	2.78	1.043	1	4	2.16	0.738
Invaded personal space	1	4	2.55	1.005	1	4	2.41	0.859
Poor customer service	1	4	1.74	0.882	1	4	1.65	0.735
Could not fully immerse into the experience	1	4	2.26	0.910	1	4	1.92	0.848
Could not see the attraction well	1	4	2.10	0.936	1	4	1.96	0.813
Could not visit the attraction properly	1	4	2.14	0.951	1	4	1.83	0.774
Could not snap the desired picture	1	4	2.18	1.053	1	4	2.21	0.927
Could not 'enjoy the moment'	1	4	2.21	1.040	1	4	1.85	0.799

Correlations

A bivariate correlation was run to ascertain the potential relationships between the variables. The results are displayed in Table 3.3. For Great Smoky Mountains, 'Overcrowding' had a significant positive correlation with 'Impact on tourist experience' (0.327). Conversely, there was no significant correlation between the 'Overcrowding and 'Impact on tourist experience' for Las Vegas. For Great Smoky Mountains, 'Type of impact on tourist experience' had significant correlations with 'Impact on tourist experience' ($-.616$), 'revisit intention' (0.427), 'trip satisfaction' (0.410) and 'trip memorability' (0.335). Similarly, for Las Vegas, 'Type of impact on experience' showed significant correlations with 'Impact on tourist experience' ($-.714$), 'revisit intention' (0.460), 'trip satisfaction' (0.357) and 'trip memorability' (0.366). Additionally, for Las Vegas, 'Impact on tourist experience' also showed a significant negative correlation with 'crowd tolerance' (-0.403) indicating crowd tolerance as a factor that can impact tourist impact at destinations. For Las Vegas, 'Impact on Tourist experience' was found to have a significant positive correlation with 'trip satisfaction', 'trip memorability' and 'crowd tolerance'. 'Impact on Tourist experience' did not have a significant correlation with 'crowd tolerance' for the Great Smoky Mountains sample.

T-Test

An independent sample t-test for equality of means was conducted to examine the differences between the two destinations. The test findings indicate a higher crowd tolerance in the Las Vegas destination (M = 1.84 as compared to 2.18 of Great Smoky Mountains) and the difference in the means was significant (F stat = 3.549, p <0.001). Further, tests were conducted to test the difference between the means of the 'Type of impact on tourist experience' (positive, negative or no impact) between the two destinations (Ross & Willson, 2017). The study tested the assumption that the difference in destination character (Vegas vs. Great Smoky Mountains) affects the overcrowding's impact on the tourist experience differently. The findings proved the assumption as the results between two groups were statistically significant (F stat = 12.790, p <0.001). The mean of 'Type of impact on tourist experience' for Las Vegas destination (M = 1.90) was higher than that of the Great Smoky Mountains group (M = 1.36). Additionally, the study also assessed the differences in the means of the 'Impact on tourist experience' (high/low/no impact on the quality of tourist experience) between the two destinations. Again, the difference in the means was statistically significant (F stat = 5.213, p <0.001) and the mean for Great Smoky Mountains was higher (M = 4.11) than Las Vegas (M = 3.12).

TABLE 3.3 Correlations

	Overcrowding	Type of Impact on Experience	Impact on Tourist Experience	Revisit Intention	Trip Satisfaction	Trip Memorability	Crowd Tolerance
Destination: Great Smoky Mountains							
Overcrowding	1						
Type of impact on experience	−0.197	1					
Impact on tourist experience	.327**	−.616**	1				
Revisit intention	−0.027	.427**	−.432**	1			
Trip satisfaction	−0.173	.410**	−.470**	.638**	1		
Trip Memorability	0.029	.335**	−.255*	.572**	.715**	1	
Crowd tolerance	−0.065	−0.041	0.140	−0.224	−0.117	−0.114	1
Destination: Las Vegas							
Overcrowding	1						
Type of impact on experience	−0.048	1					
Impact on tourist experience	0.002	−.714**	1				
Revisit intention	0.118	.460**	−.663**	1			
Trip satisfaction	0.120	.357**	−.464**	.540**	1		
Trip Memorability	0.065	.366**	−.525**	.572**	.704**	1	
Crowd tolerance	−0.113	−.403**	.435**	−.214*	−.346**	−.320**	1

* Correlation is significant at the 0.05 level (2-tailed).
** Correlation is significant at the 0.01 level (2-tailed).

Regression

Given the correlation between 'Overcrowding', 'Type of impact on tourist experience' and 'Impact on tourist experience', a multiple regression was employed to examine the relationship between the variables. The results are presented in Table 3.4. 'Overcrowding' had a significant positive relation with 'Impact on tourist experience' ($\beta = 0.768$, t = 2.160, p<0.05), while controlling 'crowd tolerance' and 'type of tourist experience'. The findings show that overcrowding negatively influences tourists experience at the destination. Additionally, increased 'Crowd tolerance' also has a positive influence ($\beta = 0.524$, t = 2.799, p<0.01) on 'tourist experience' when controlled for 'crowded attractions' and 'type of impact on tourist experience'. Conversely, the 'type of impact on tourist experience' had a significant negative relation ($\beta = -0.700$, t = -3.933, p<0.001) when controlling for number of crowed places and crowd tolerance. This finding further supports the hypothesis that with the increase in tolerance

TABLE 3.4 Regression results

Relationships tested	B	S.E.	t-value	p-value	Supported?
Overcrowding > Impact on tourist experience	0.768	0.355	2.160	0.032	Yes
Type of impact > Impact on tourist experience	−0.700	0.178	−3.933	0.000	Yes
Overcrowding > Trip satisfaction	0.064	0.083	0.771	0.442	No
Overcrowding > Trip memorability	0.106	0.064	1.669	0.097	No
Overcrowding > Revisit intention	0.198	0.101	1.958	0.052	No
Crowd tolerance > Impact on tourist experience	0.524	0.187	2.799	0.006	Yes
Impact on tourist experience > Trip satisfaction	−0.270	0.088	−3.079	0.002	Yes
Impact on tourist experience > Trip memorability	−0.050	0.069	−0.729	0.467	No
Impact on tourist experience > Revisit intention	−0.350	0.107	−3.282	0.001	Yes
Trip satisfaction > Revisit intention	0.382	0.127	2.998	0.003	Yes
Trip memorability > Revisit intention	0.387	0.128	3.018	0.003	Yes
Moderating Effects					
Relationships tested					
Overcrowding > Type of impact > Impact on tourist experience	−0.165	0.131	−1.260	0.209	No
Crowd tolerance > Destination character > Impact on tourist experience	0.578	0.228	2.534	0.012	Yes

for crowds, the perceptions of crowding vary and the any negative influence on destination experience is reduced.

Multiple regression

Given the significant correlation between tourist experience and trip satisfaction, trip memorability and revisit intention, a set of multiple regressions were conducted.

First, we tested the relationship between number of overcrowded places (overcrowding), crowd tolerance, type of tourist experience, impact on tourist experience and trip satisfaction. The results indicate that overcrowding's level of 'impact on tourist experience' has a significant negative influence on 'trip satisfaction' ($\beta = -0.270$, $t = -3.079$, $p<0.01$) when controlled for crowd tolerance, number of overcrowded places and type of tourist experience, meaning, the higher the (negative) impact of overcrowding on tourist experience, the lower the trip satisfaction.

A second set of multiple regression tested the impact of number of overcrowded places (overcrowding), crowd tolerance, type of tourist experience, impact on tourist experience, and trip satisfaction on trip memorability. The findings showed that trip satisfaction positively influenced trip memorability ($\beta = 0.649$, $t = 10.657$, $p<0.001$) when controlling for the number of crowded attractions, crowd tolerance, type of tourist experience, impact on tourist experience and trip satisfaction. Thus, it can be inferred that tourists that are satisfied with the trip will form positive memories of the trip.

The last set of multiple regression tested the impact of number of overcrowded places (overcrowding), crowd tolerance, type of tourist experience, impact on tourist experience, trip satisfaction and trip memorability on revisit intention. The results indicate that overcrowding's level of (negative) 'impact on tourist experience' has a significant negative effect ($\beta = -0.368$, $t = -3.359$, $p<0.001$) on 'revisit intention' when controlling for number of crowded places, crowd tolerance, trip satisfaction, trip memorability, the type of impact on tourist experience. Meaning, the higher the (negative) impact of overcrowding on tourist experience, the lower the revisit intention.

Both trip satisfaction ($\beta = 0.386$, $t = 3.022$, $p<0.01$) and trip memorability ($\beta = 0.388$, $t = 3.025$, $p<0.01$) have a significant positive effect on revisit intention when controlling for other variables in the equation.

Discussion

The chapter aimed to investigate the impact of overcrowding on tourist experience in diverse destination contexts. While Las Vegas is globally well known as a gambling, entertainment and conference destination visited by large crowds all year around, the destination of the Gatlinburg is the prime getaway community for the Great Smoky Mountains National Park, the most visited national park in the United States.

The destination character of Las Vegas is a tourism epicentre with one major centre, and the tourist destination of the Great Smoky Mountains/ Gatlinburg is a popular destination with a multi-centre character, i.e., multiple tourist attractions are spread out along the county, including the national park visitor centre and major trails, Dollywood theme park in Pigeon Forge, and several outlet malls for shopping. Meanwhile, the destination has a nature-based, country-like character with a lot of cultural, nature- and activity-based attractions. It allows the visitor numbers to distribute better in space centring around multiple tourist attractions based on their interest and motivation, however, crowding or even overcrowding at certain attractions is still a common phenomenon.

Crowd tolerance is shaped by the destination's character

The Vegas respondents showed a slightly higher propensity to tolerate crowds (26.74% said they did not mind crowds as opposed to only 12.5% of the Great Smoky Mountains respondents). On average, crowd tolerance for Las Vegas visitors was higher than that of Great Smoky Mountains visitors and the mean difference was significant.

The findings of this study confirmed that destination character is a major moderating factor in the relationship between crowd tolerance (independent variable) and the level of overcrowding's negative impact on the tourist experience (dependent variable). The findings contribute to the body of knowledge by adding destination character to the list of factors impacting crowd tolerance – personal traits (Kim and Yoon, 2020), tourist profile (Jurado et al., 2013), consumed tourism product (Jacobsen et al., 2019).

Destination character determines the impact of overcrowding on tourist experience

In the case of Great Smoky Mountains, experienced overcrowding ('Overcrowding') had a significant correlation with its negative impact on tourist experience ('Tourist experience'). Conversely, there was no significant correlation between the experienced overcrowding and the impact of overcrowding on tourist experience for Las Vegas, i.e., higher levels of experienced overcrowding do not necessary lead to perceived negative impact on tourist experience.

The study found that Great Smoky Mountains visitors had a higher incidence of negative experiences as compared to Las Vegas visitors. All in all, Great Smoky Mountains had higher mean scores on negative impacts of overcrowding on tourist experience than Las Vegas. This includes areas such as long wait times, invaded personal space, poor customer service, limited immersion into the experience, limited enjoyment of the moment, and visiting or seeing the attraction well. The only area where Las Vegas performed worse in regard to overcrowding's negative impact on the tourist experience than Great Smoky Mountain is limitation on snapping the desired picture.

The reasons why crowding impacts the tourist experience differently in destinations with distinct character are numerous. Factors such as anticipated experiences and tourist motivation shape crowd tolerance (especially crowd tolerance in-situ), therefore, the coping mechanism to overcrowding vary in different type of destinations. Papadopoulou et al. (2023), for instance, point out that visitors might be able to develop positive coping mechanisms to crowding and become more resilient to such experiences having less of an effect on their intention to revisit or recommend the destination. Our findings extend the findings of their study suggesting that coping mechanisms with overcrowding is also determined by destination's character.

Several researches on crowding have been conducted in natural parks and protected areas, where tourists seek to enjoy nature. Consequently, the ideal condition in these studies is often seen as having zero crowding, with a sharply negative norm curve (Vaske & Shelby, 2008; Rathnayake, 2015). The current study confirmed a significant difference in overcrowding's negative impact on tourist experience in the two studied destinations.

The level of overcrowding has an indirect impact on trip satisfaction and revisit intention

Overcrowding's impact on tourist experience was perceived by visitors as more negative in the case of Great Smoky Mountains than Las Vegas. This impact was measured via two different measurement items ('Type of impact on experience' and 'Tourist experience') and both results confirmed this finding. Based on the findings, the higher the (negative) impact of overcrowding on tourist experience, the lower is the trip satisfaction and the revisit intention.

Several previous researches argue that higher perceptions of crowding tend to evoke a more negative response, leading to lower levels of satisfaction (e.g., Neuts & Nijkamp, 2012; Rathnayake, 2015; Zhang et al., 2017). Although the findings of this study might concur with these previous studies at first sight, but when looking closer, we can see that the impact of the level of overcrowding ('number of overcrowded places') is indirect in determining tourist satisfaction and revisit intention (no significant direct impact). The impact of overcrowding on tourist experience ('impact on experience') is variable with a direct impact on tourist satisfaction and revisit intention. Meaning, rather than the level of overcrowding, the perceived impact of overcrowding on the visitor experience is what matters. The findings of the current study point out that the impact of overcrowding has different dynamics with different results in various destination contexts.

Positive crowding experience is more likely in city-like tourism hotspots

Destination experience due to overcrowding is a critical aspect of tourism development of the destination. Social interference theory has important implications for understanding how social cues in the tourism environment can impact cognitive performance and decision-making. In the tourism context, social cues can include interactions with other tourists, the behaviour and actions of locals, and the general atmosphere of the destination. Positive crowding perceptions also exist in the literature (Neuts and Nijkamp 2012; Popp 2012). Marketing literature suggests that crowds enhance the experiences in hedonic services such as amusement parks or concerts and retail settings (Pons et al., 2006). A number of experts (Pons, Laroche, and Mourali 2006; Sun and Budruk 2017; Tokarchuk et al., 2022) agree that crowds can add value to an experience. Popp (2012) proposed that good crowding, during a high-density visit, might occur due to human-related motivations, for example sharing experiences with people, socialising, and observing others, or undertaking group activities. Naturally, overcrowding perception varies with individuals, but it is also dependent on the type of tourism developed in the area, i.e., destination character. Thus, tourists visiting Disneyland or New York are expected to be more tolerant to (or attracted by) crowding than people who travel to other destinations offering attractions related to natural resources, such as beaches or forests (Santana-Jimenez and Hernandez, 2011).

Negative crowding experience is more likely in nature-based leisure destinations

Nature-based leisure destinations seem to have significantly more negative tourist experience due to crowded attractions as compared to city-like tourist hotspot destinations.

The current study shows that visitors seem to anticipate less crowds in nature-based leisure destinations when compared to major tourism epicentres. This may be due to the additional cognitive demands of processing the social cues and behaviour of other tourists, leaving fewer resources available for enjoyment and engagement with the destination experience. These types of destinations are rather based on experience and enjoyment of natural and cultural resources than entertainment or business. In other words, enjoying the moment or getting into a flow state of mind in a context that is not dependent on the social vibe of the destination. At the same time, some previous findings suggest that overcrowding at tourist attractions detracts from the experience (Tseng et al., 2009). Tourists in smaller, but popular, destinations like Great Smoky Mountains/Great Smoky Mountains found their experience (negatively) impacted by overcrowding in larger extent and showed lower intentions to revisit the destination.

> As previously mentioned, we've been visiting for 15 years. I love Pigeon Forge and Gatlinburg. However, every year, the more crowded it becomes and the less pleasurable it continues to get AND for the price of the experiences knowing it cannot be fully enjoyed, we have considered vacationing elsewhere.
>
> *(Great Smoky Mountain respondent, no. 54)*

> Too many folks to enjoy for the amount of time I wanted too.
>
> *(Great Smoky Mountain respondent, no. 22)*

> Could not do any of the activities we wanted to do due to too many people.
>
> *(Great Smoky Mountain respondent, no. 67)*

Crowding can enhance the authenticity of the tourist experience by creating a sense of place and identity

In some cases, crowding or even overcrowding can enhance the authenticity of the tourist experience by creating a sense of place and identity. This is especially true when overcrowding is perceived as a natural consequence of the destination's popularity and not the result of excessive tourism development. This is the case in Las Vegas. As a tourism hotspot globally well known as the number one destination for entertainment and gambling, a lively vibe described by big crowds very much defines the sense of place of the destination. Visitors expect crowding while visiting the destination, part of it given by the mega size of the hotels, resorts, arenas and convention centres. The entertainment industry based on large-scale casinos and spectacular shows was designed for crowds to gather and celebrate together. Vegas would not be the same without crowds. Given the anticipated experience, the visitors coping mechanism with overcrowding improves (see Papadopolou et al., 2023) resulting in better outcome on the overcrowding's impact on the tourist experience.

Tokarchuk et al. (2022) argue that larger crowds in the destination attract visitors who enjoy crowds over time, thus evolving to a mass destination. Their study on Berlin points out positive effects of crowding in highly experiential contexts wherein tourists primarily seek enjoyment and emotional value (Pons et al., 2006). The current study found that visitors in Las Vegas showed significantly higher crowd tolerance than their counterparts in Great Smoky Mountains and found the crowds to be a positive cognitive experience. The visitors believed that crowds added to their overall experience as they added to the vibe of the place. The respondents of the survey had a chance to share their experience in the form of open-ended questions:

I was expecting the overcrowding when I visited LV and that was one of my expectations as a tourist. Actually, this overcrowding provided a more immersive experience that I truly felt that I am in this one of the most popular tourist destinations that is well known for crowded, exciting, and stimulating.

(Las Vegas respondent no. 25)

Crowds are the essence of tourist places like Las Vegas. I wouldn't like it to be 'deserted'.

(Las Vegas respondent no. 7)

Crowding's positive effects on the tourist experience include creating a sense of excitement and social interaction

Crowding can contribute to creating a sense of excitement and social interaction during the visit. However, this is only true when crowding is perceived as manageable and does not lead to negative consequences such as congestion, pollution and degradation of the environment. This is especially valid in the case of more nature-based, smaller destinations (i.e., less of a big city like destinations) such as Great Smoky Mountains. Although crowds can elevate the sense of excitement and can amplify social interaction, the visitors' crowd tolerance is lower when visiting such destinations. Their anticipated experience does not necessarily include dealing with inconveniences caused by crowding, such as long waiting times or not being able to enjoy the experience more quietly (for example, in a park).

The overcrowding did not affect my experience because I really enjoyed interacting with other people of course I couldn't get other services very fast, but it wasn't really a bad experience I enjoyed every moment.

(Las Vegas respondent no. 38)

I think the casino was fun because it had people. I'm not really an introvert so I like crowded places.

(Las Vegas respondent no. 58)

Lots of fun having many people around all gambling on the same table.

(Las Vegas respondent no. 13)

Conclusion

The findings of the study uncover several new aspects within the relationship dynamics of destination character, level of overcrowding, crowd tolerance and the impact of overcrowding on visitor experience, namely:

- Crowd tolerance is shaped by the destination's character.
- Destination character determines the impact of overcrowding on tourist experience.
- The level of overcrowding has an indirect impact on trip satisfaction and revisit intention.
- Positive crowding experience is more likely in city-like tourism hotspots.
- Negative crowding experience is more likely in nature-based leisure destinations.
- Crowding can enhance the authenticity of the tourist experience by creating a sense of place and identity.
- Crowding's positive effects on the tourist experience include creating a sense of excitement and social interaction.

The study contributes to the existing literature on overcrowding and its impact on the tourist experience in destinations with distinct characters. Despite the limitations of the study, these findings represent valuable insights to the theoretical and managerial discussions on tourism overcrowding.

Managerial implications

While overcrowding is generally considered a negative phenomenon in tourism, it is important to consider the context and the perception of the tourist experience. By understanding the factors that contribute to the positive and negative impacts of overcrowding, tourism providers can develop strategies to manage tourism flows and enhance the quality of the tourist experience.

There is no one-size-fits-all solution for overcrowding. Crowding and, more specifically, overcrowding manifest themselves in multiple ways depending on the context of different tourism stakeholders and different destination types.

As Koens et al. (2018) point out, the way in which overtourism manifests itself, and the possibilities for dealing with the issues arising from overcrowding, strongly depend on the destination context. This study adds to the current discussion by highlighting the role of destination character in regard to its role shaping overcrowding's impact on tourist experience. It is suggested that addressing overcrowding solutions needs to be made to fit the destination specific context whether it relates to managing the impacts of overcrowding or preventing overcrowding.

Much of the discussion regarding overcrowding focuses on the tourist epicentres, however, it is not just an issue in cities, but it can also be observed in smaller-scale and nature-based destinations. Considering and discussing the role of the destination character on all aspects of overcrowding will enrich the debate on issues related to crowding at a destination level. It is recommended that industry experts, practitioners and academic researchers continue to engage with the issue.

Limitations and future research directions

Overcrowding has been an increasing issue within many tourist destinations, both pre- and post-COVID-19 pandemic, prompting increased research interest in this area. Despite the above contributions, the study has some limitations that may be addressed by future research.

First, this study had limited sample size; thus, the results should be interpreted with caution. The study is to be viewed as exploratory in its nature, whereas the goal was rather working towards model building than testing. Consequently, during the data analysis only less-complex methods were applied.

Second, this study was undertaken at two popular destinations in the United States, and the respondents were predominantly domestic visitors. Future studies may include destinations of different characters or outside the United States for further analysis. Future studies may also undertake an analysis of the differences between domestic and international visitors as well as differences between similar destinations, i.e., sharing a similar destination character, in various countries.

Finally, although this study considers crowd tolerance as one of the factors that may impact the tourist experience due to overcrowding, it does not account for the cultural differences which may impact the crowd tolerance of the visitors. This represents a gap in literature that would be worth the investigation.

References

Bloom, L. (2019). Bucket list travel: The top 50 places in the world. Forbes. https://www.forbes.com/sites/laurabegleybloom/2019/09/04/bucket-list-travel-the-top-50-places-in-the-world/?sh=fdf020620cfd

Butler, R. (2018). Challenges and opportunities. *Worldwide Hospitality and Tourism Themes*, 10(6), 635–641. https://doi.org/10.1108/WHATT-07-2018-0042

Capocchi, A., Vallone, C., Amaduzzi, A., & Pierotti, M. (2019). Is "overtourism" a new issue in tourism development or just a new term for an already known phenomenon? *Current Issues in Tourism*. https://doi.org/10.1080/13683500.2019.1638353

Choi, S. C., Mirjafari, A., & Weaver, H. B. (1976). The concept of crowding: A critical review and proposal of an alternative approach. *Environment and Behavior*, 8(3), 345–362. https://doi.org/10.1177/136327527600800302

Cranley, E. (2019, August 3). *11 places that Instagram made famous*. Insider. https://www.insider.com/instagram-travel-tourists-popular-places-nature-cities-2019-8

Dodds, R. & Butler, R. (2019). *Overtourism: Issues, realities and solutions*. De Gruyter Oldenbourg. https://doi.org/10.1515/9783110607369

Douglass, W., & Raento, P. (2004). The tradition of invention: Conceiving Las Vegas. *Annals of Tourism Research*, 31, 7–23.

Emerson, R. M. (1976). Social exchange theory. *Annual Review of Sociology*, 2, 335–362. https://doi.org/10.1146/annurev.so.02.080176.002003

Fishman, J. (2015, December 31). TripAdvisor names city of Great Smoky Mountains the nation's #1 destination on the rise. Sevier County, TN: Visit My Smokies https://www.visitmysmokies.com/blog/GreatSmokyMountains/city-of-GreatSmoky Mountains-is-top-destination-on-the-rise/

Francis, J. (2023). Overtourism mapped: Tourism is headed into a global crisis. Responsible Travel. https://www.responsiblevacation.com/copy/overtourism-map

Gan, V. (2015, August 18). *Which US cities and states hate tourists the most?* CitiLab. https://www.bloomberg.com/news/articles/2015-08-28/which-u-s-cities-and-states-hate-tourists-the-most-according-to-a-study-by-stratos-jets

Goodwin, H. (2011). *Taking responsibility for tourism*. Goodfellow Publishers Ltd. Retrieved from https://www.perlego.com/book/868824/taking-responsibility-for-tourism-pdf (Original work published 2011)

Goodwin, H. (2017). The challenge of overtourism. *Responsible Tourism Partnership*, 4, 1–19.

Gössling, S., McCabe, S., & Chen, N. C. (2020). A socio-psychological conceptualisation of overtourism. *Annals of Tourism Research*, 84, 102976. https://doi.org/10.1016/j.annals.2020.102976

Hootboard (2023). How DMOs Can Manage Overtourism? By Bryan Reynolds, Destination Marketing, April 4, 2023. https://about.hootboard.com/destination-marketing/how-dmos-can-manage-overtourism/

Insch, A. (2020). The Challenges of Over-tourism Facing New Zealand: Risks and responses, *Journal of Destination Marketing & Management*, 15, https://doi.org/10.1016/j.jdmm.2019.100378

Jacobsen, J. K. S., Iversen, N. M., & Hem, L. E. (2019). Hotspot crowding and over-tourism: Antecedents of destination attractiveness. *Annals of Tourism Research*, 76, 53–66. https://doi.org/10.1016/j.annals.2019.02.011

Jurado, E. N., Damian, I. M., & Fernandez-Morales, A. (2013). Carrying capacity model applied in coastal destinations. *Annals of Tourism Research*, 43, 1–19. https://doi.org/10.1016/j.annals.2013.03.005

Kainzinger, S., Burns, R. & Arnberger, A. (2015) Whitewater boater and angler conflict, crowding and satisfaction on the North Umpqua River, Oregon. *Human Dimensions of Wildlife*, 20, 542–552. https://doi.org/10.1080/10871209.2015.1072757

Kim, H.-R., & Yoon, S.-Y. (2020). How to help crowded destinations: Tourist anger vs. sympathy and role of destination social responsibility. *Sustainability*, 12(6), 2358. https://doi.org/10.3390/su12062358

Kock, F., Berbekova, A., & Assaf, A. G. (2021). Understanding and managing the threat of common method bias: Detection, prevention and control. *Tourism Management*, 86, 104330. https://doi.org/10.1016/j.tourman.2021.104330

Koens, K., Postma, A., & Papp, B. (2018). Is overtourism overused? Understanding the impact of tourism in a city context. *Sustainability*, 10, 4384. https://doi.org/10.3390/su10124384

Kwon, H., Ha, S., & Im, H. (2016). The impact of perceived similarity to other customers on shopping mall satisfaction. *Journal of Retailing and Consumer Services*, 28, 304–309. https://doi.org/10.1016/j.jretconser.2015.01.004

Lee, J., Benjamin, S., & Childs, M. (2020). Unpacking the emotions behind TripAdvisor travel reviews: The case study of great smoky mountains, Tennessee. *International Journal of Hospitality & Tourism Administration 2022*, 23(10), 1–18. https://doi.org/10.1080/15256480.2020.1746219

Li, L., Zhang, J., Nian, S., & Zhang, H. (2017). Tourists' perceptions of crowding, attractiveness, and satisfaction: a second-order structural model. *Asia Pacific Journal of Tourism Research*, 22, 1250–1260. https://doi.org/10.1080/10941665.2017.1391305

Li, T. T., Liu, F., & Soutar, G. N. (2021). Experiences, post-trip destination image, satisfaction and loyalty: A study in an ecotourism context. *Journal of Destination Marketing & Management*, 19, 100547. https://doi.org/10.1016/j.jdmm.2020.100547

LVCVA. (2023, April 11). Economic impact driven by Las Vegas tourism industry hits record high $79.3 Billion in 2022. Las Vegas Convention and Visitors Authority. https://press.lvcva.com/news-releases/economic-impact-las-vegas-tourism-industry-hits-record-high-79.3-billion-2022/s/09510b4a-a9d6-4180-9aec-9cd8c385b347

Machleit, K. A., Eroglu, S. A., & Mantel, S. P. (2000). Perceived retail crowding and shopping satisfaction: What modifies this relationship? *Journal of Consumer Psychology*, 9(1), 29–42. https://doi.org/10.1207/s15327663jcp0901_3

McKinsey & Company. (2017). Coping with success: Managing overcrowding in tourism destinations. https://www.mckinsey.com/~/media/mckinsey/industries/travel%20logistics%20and%20infrastructure/our%20insights/coping%20with%20success%20managing%20overcrowding%20in%20tourism%20destinations/coping-with-success-managing-overcrowding-in-tourism-destinations.pdf

Mihalic, T. (2020). Conceptualising overtourism: A sustainability approach. *Annals of Tourism Research*, 84. https://doi.org/10.1016/j.annals.2020.103025

Milano, C., Novelli, M., & Cheer, J. M. (2019). Overtourism and degrowth: A social movements perspective. *Journal of Sustainable Tourism*. https://doi.org/10.1080/09669582.2019.1650054

Nelson, L. 2018. *How are New Orleans and Las Vegas coping with overtourism?* Griffith Institute for Tourism: Tourism Dashboard. https://www.tourismdashboard.org/how-are-new-orleans-and-las-vegas-coping-with-overtourism/

Namberger P., Jackisch S., Schmude J., & Karl M. (2019). Overcrowding, overtourism and local level disturbance: How much can Munich handle? *Tourism Planning & Development*, 16(4), 452–472. https://doi.org/10.1080/21568316.2019.1595706

National Park Service (2023). *Park Statistics.* https://www.nps.gov/grsm/learn/management/statistics.htm

Neuts, B., & Nijkamp, P. (2012). Tourist crowding perception and acceptability in cities: An applied modelling study on Bruges. *Annals of Tourism Research*, 39(4), 2133–2153. https://doi.org/10.1016/j.annals.2012.07.016

Oklevik, O., Gössling, S., Hall, C. M., Steen Jacobsen, J. K., Grøtte, I. P., & McCabe, S. (2019). Overtourism, optimisation, and destination performance indicators: A case study of activities in Fjord Norway. *Journal of Sustainable Tourism*, 27(12), 1804–1824. https://doi.org/10.1080/09669582.2018.1533020

Papadopoulou, N. M., Ribeiro, M. A., & Prayag, G. (2023). Psychological determinants of tourist satisfaction and destination loyalty: The influence of perceived overcrowding and over tourism. *Journal of Travel Research*, 62(3), 644–662. https://doi.org/10.1177/00472875221089049

Peters, P., Gosslin, S., Milano, C., Novelli, M., Dijkmans, C., Eijgelaar, E., Hartman, S., Heslinga, J., Isaac, R., Mitas, O., Moretti, S., Nawijn, J., Papp, B., & Postma, A. (2018, October). Research for TRAN committee - Overtourism: Impact and possible policy responses. European Parliament, Policy Department for Structural and Cohesion Policies, Brussels. http://www.europarl.europa.eu/RegData/etudes/STUD/2018/629184/IPOL_STU(2018)629184_EN.pdf

Peltier, D. (2018). U.S. National Parks still aren't sure how to deal with overtourism. Skift. https://skift.com/2018/03/02/u-s-national-parks-arent-sure-how-to-deal-with-overtourism/

Perdue, R. R., Long, P. T., & Kang, Y. S. (1999). Boomtown tourism and resident quality of life: The marketing of gaming to host community residents. *Journal of Business Research*, 44(3), 165–177. https://doi.org/10.1016/S0148-2963(97)00198-7

Podsakoff, P. M., MacKenzie, S. B., & Podsakoff, N. P. (2012). Sources of method bias in social science research and recommendations on how to control it. *Annual Review of Psychology*, 63(1), 539–569. https://doi.org/10.1146/annurev-psych-120710-100452

Pons, F., & Laroche, M. (2007). Cross-cultural differences in crowd assessment. *Journal of Business Research*, 60(3), 269–276. https://doi.org/10.1016/j.jbusres.2006.10.017

Popp, M. (2012). Positive and negative urban tourist crowding: Florence, Italy. *Tourism Geographies*, 14(1), 50–72. https://doi.org/10.1080/14616688.2011.597421

Ramkissoon, H., Wong, J. W. C., Zhou, Z., & Xue, J. (2022). Stimulating tourist inspiration by tourist experience: The moderating role of destination familiarity. *Frontiers in Psychology*, 13, 895136. https://doi.org/10.3389/fpsyg.2022.895136

Rathnayake, R. M. W. (2015). How does 'crowding' affect visitor satisfaction at the Horton Plains National Park in Sri Lanka? *Tourism Management Perspectives*, 16, 129–138. https://doi.org/10.1016/j.tmp.2015.07.018

Ross, A., & Willson, V.L. (2017). *Independent Samples T-Test. In: Basic and Advanced Statistical Tests.* SensePublishers. https://doi.org/10.1007/978-94-6351-086-8_3

Salem, M. (2022, September 11). *How many casinos are in Las Vegas?* Best Hotel for You. https://www.besthotelfor.com/how-many-casinos-are-in-las-vegas/

Santana Jiménez, Y., & Hernández, J. M. (2011). Estimating the effect of overcrowding on tourist attractions: The case of Canary Islands. *Tourism Management*, 32(2), 415–425. https://doi.org/10.1016/j.tourman.2010.03.013

Shelby, B., Vaske, J. J., & Heberlein, T. A. (1989). Comparative analysis of crowding in multiple locations: Results from fifteen years of research. *Leisure Sciences*, 11(4), 269–291. https://doi.org/10.1080/01490408909512227

Spanier, D. (1992). *Welcome to the pleasure dome: Inside Las Vegas.* University of Nevada Press.

Stokols, D. (1972). A social-psychological model of human crowding phenomena. *Journal of the American Institute of Planners,* 38(2), 72–83. https://doi.org/10.1080/01944367208977409

Sun, Y. Y., & M. Budruk. (2017). The moderating effect of nationality on crowding perception, its antecedents, and coping behaviours: A study of an urban heritage site in Taiwan. *Current Issues in Tourism,* 20(12), 1246–1264. https://doi.org/10.1080/13683500.2015.1089845

TasteAtlas. (2023, September 20). What happened to Portofino? A cautious tale of losing authenticity. *World Food Atlas: Discover 17060 Local Dishes & Ingredients.* https://www.tasteatlas.com/what-happened-to-portofino-a-cautious-tale-of-losing-authenticity

Tokarchuk, O., Barr, J.C., & Cozzio, C. (2022). How much is too much? Estimating tourism carrying capacity in urban context using sentiment analysis. *Tourism Management,* 91. https://doi.org/10.1016/j.tourman.2022.104522

Tooman, L.A. (1997). Tourism and development. *Journal of Travel Research* (Winter 1997), 33–40. https://doi.org/10.1177/004728759703500305

TripAdvisor. (2016, September 1). 24 beautiful little mountain towns across America. TripAdvisor.com https://wwwtripadvisorcom/VacationRentalsBlog/2016/09/01/beautiful-little-mountaintowns-across-america/

Tseng, Y., Kyle, G.T., Shafer, C.S., Graefe, A.R., Bradle, T.A., & Schuett, M.A. (2009). Exploring the crowding–satisfaction relationship in recreational boating. *Environmental Management,* 43, 496–507. https://doi.org/10.1007/s00267-008-9249-5

Turkle, S. (2015). *Reclaiming conversation.* Basic Books.

Turner, L., & Ash, J. (1975). *The golden hordes: International tourism and the pleasure periphery.* Constable.

Vaske, J. J., & Shelby, L. B. (2008). Crowding as a descriptive indicator and an evaluative standard: Results from 30 years of research. *Leisure Sciences,* 30(2), 111–126. https://doi.org/10.1080/01490400701881341

Vlamis, K. (2023, October 27). Residents of the stunning Austrian village rumored to have inspired "frozen" are fed up with hoards of tourists. *Insider.* https://www.insider.com/hallstatt-austria-locals-protest-overtourism-frozen-village-photos-2023-10

Wall, G. (2020). From carrying capacity to overtourism: A perspective article. *Tourism Review,* 75(1). https://doi.org/10.1108/TR-08-2019-0356

Warrick, M. (2022, July 30). *Dollywood pushing big bucks into TN economy.* WVLT Knoxville. https://www.wvlt.tv/2022/07/30/dollywood-pushing-big-bucks-into-tn-economy/

Weaver, D. (2011). Contemporary tourism heritage as heritage tourism: Evidence from Las Vegas and Gold Coast. *Annals of Tourism Research,* 38(1), 249–267. https://doi.org/10.1016/j.annals.2010.08.007

Weber, F., Stettler, J., Priskin, J., Rosenberg-Taufer, B., Ponnapureddy, S., Fux, S., Camp, M. A., Barth, M., Klemmer, L., & Gross, S. (2017). Tourism destinations under pressure. Working Paper, Institute of Tourism ITW, Lucerne University of Applied Sciences and Arts. https://static1.squarespace.com/static/56dacbc6d210b821510cf939/t/5906f320f7e0ab75891c6e65/1493627704590/WTFL_study+2017_full+version.pdf

Wood, A. (2005). "What happens [in Vegas]": Performing the post-tourist flaneur in "New York" and "Paris". *Text and Performance Quarterly*, 25(4), 315–333. https://doi.org/10.1080/10462930500362403

Yu J., Egger R. (2020). Tourist experiences at overcrowded attractions: A text analytics approach. *Information and Communication Technologies in Tourism*, 2021 November 28, 231–243. https://doi.org/10.1007/978-3-030-65785-7_21

Zhang, Y., Li, X. R., Su, Q., & Hu, X. (2017). Exploring a theme park's tourism carrying capacity: A demand-side analysis. *Tourism Management*, 59, 564–578. https://doi.org/10.1016/j.tourman.2016.08.019

4

FROM TRUST TO TRUSTWORTHINESS

Formalising consumer behaviour with discourse on Airbnb platform

Aurimas Pumputis and Micol Mieli

Introduction

Online platforms that appeared with the prominence of the sharing economy are widely used by consumers in tourism for gaining access to assets, opinions and services, including touristic accommodation. Accommodation as an asset shared on platforms requires connecting consumers with service providers that are individual people unknown to the consumer. Transactions with them become risky for tourists due to the need to book services far from home, in places where the consumer lacks tacit knowledge of operating in that context (Williams & Baláž, 2015). As consumers, tourists that use P2P (peer-to-peer) platforms are displaced and have to rely on unknown peers as service providers, which creates risks unique to the use of online platforms (Wirtz et al., 2019). They may include unethical behaviour of the service provider (Wirtz et al., 2019), privacy infringements (Teubner & Flath, 2019), or simply poor service quality that would harm the consumer's plans (Huang et al., 2020). In one-off encounters with unknown peers, the consumers need to carefully weigh whether the benefits of using a platform outweigh such risks.

A particularly prominent platform where such risks are noticed as a source of discontinuing the platform's use is Airbnb (Huang et al., 2020). Numerous studies have explored how Airbnb constructs trust between consumers and providers via its user-reviews based online reputation management system (e.g., Baute-Díaz et al., 2019; Ert et al., 2016). Trust, which can be fostered with the help of the platform's design, is a powerful tool to mitigate previously discussed risks. However, in the setting of P2P platforms, trust can be differentiated between trust in the platform, and interpersonal trust between consumers and service providers. As demonstrated by Pelgander et al. (2022), trust in the

DOI: 10.4324/9781032637778-5

platform is based on its functions but may not involve an emotional commitment as seen in interpersonal relations. Building trust between consumers and providers on the basis of online reviews is among the platform's functions. Online reviews help establish a basis for trust, sufficient to enter an economic transaction; however, real trust is built in interactions, as persons repeatedly engage with each other (Korsgaard, 2018). Yet, throughout the years Airbnb has maintained a sharing and trust focused discourse about its platforms as a community marketplace (Celata et al., 2017). In this chapter, we argue that trust on platforms is also constructed through discourse in official documents such as legal terms, policies and guidelines, where the behaviour of the trustworthy consumer is formalised.

In order to understand such formalisation, we analyse the discourse presented by the platform to its users in policy documents and community guidelines as bases for trust in the 'global community' of Airbnb users (Airbnb, 2023). We argue that documents, such as the Airbnb's Terms of Use, Community Guidelines, and Community Policies, create a narrative of a 'trustworthy consumer' that is presented to Airbnb's consumer base. As a basis for trusting the platform to be a marketplace where trustworthy strangers meet, these documents also formalise consumption practices, directing consumer's behaviour towards a set standard. However, standards are not simply set for consumers to adapt to. Consumers that use Airbnb rarely read required policy documents due to their formalistic and cumbersome to understand format (Zamani et al., 2019). However, policies and terms of use are a part of a platform's design; they designate what interactions are allowed or constricted and serve as a basis when decisions are disputed. This directs consumers towards desired behaviours and creates a formal basis for already existing behaviours.

The analytical part of this chapter is based on a discourse analysis, following an understanding that discursive practices can be used to govern (Skålén et al., 2006). A discourse analysis posits the existence of different social realities in which consumers operate (Potter, 1996). In this case, we examine how discourse is used to reify what Airbnb understands as a 'trustworthy' consumer, i.e., setting a formal basis for what is considered trustworthy in consumption practices, and representing the platform's marketplace as a place where others can be trusted. This is a constructivist view, suggesting that descriptions and accounts construct the marketplace for consumers (Potter, 1996).

As recent studies about Airbnb indicate, the service providers on the platform are increasingly professionalised (Gil & Sequera, 2020), and the platform supports such development by simultaneously developing features that help provide professional hosting services. However, the rhetoric elements of sharing and trust are still widely communicated to consumers (Schor & Vallas, 2021). The limited social interaction resulting from professionalised Airbnb's supply also increases the need to employ socio-technical mechanisms for managing trust, and the need to develop specific policies that define their use

(Fraanje & Spaargaren, 2019). As we argue, these elements formalise consumption practices by codifying and regulating consumers' behaviour to correspond with the changing platforms operating model. Having established the changes of the operating model noted in previous studies, we will now move on to analysing how the standards for behaviour are created in this environment.

After reviewing the existing literature on trust in the Airbnb community and on how discourse can contribute to formalising consumer behaviour, the chapter will continue with a presentation of the methodology and the analysis of Airbnb discourse on trust, followed by a concluding discussion.

Literature review: Trust in Airbnb's community marketplace

Trust is a powerful tool and a requirement for building relationships with customers (Morgan & Hunt, 1994). Airbnb earns a revenue from matching consumers and accommodation providers in an online marketplace, where each consumer is essentially a stranger (Perren & Kozinets, 2018). Over time, it has built a distinct form of managing trust between consumers and providers, based on its discourse about building a community of travellers where 'anybody can truly belong' (Airbnb, 2023). Studies show that despite maintaining a rhetoric of community, the supply of Airbnb market is increasingly professionalised and dependent on the platform's socio-technical elements that facilitate hosting (Bosma, 2022). As a result, the platform needs to adjust its model of building trust from a peer-to-peer to a peer-to-business basis, which is different from the usual rhetoric of sharing.

This has implications for how consumers can build trust in Airbnb's service providers. In consumer behaviour theory, trust is generally linked to confidence in the exchange partner (Moorman et al., 1993) and is understood as a dynamic process of exceeding expectations in repeated interactions over time (Fam et al., 2004). However, in marketplaces such as Airbnb it is rare to engage in multiple interactions over time, and trust needs to be built on a basis of reputation (Williams & Baláž, 2021). Celata et al. (2017) further suggest that reputation is not a sufficient basis for trust in such marketplaces, and further means of increasing sociality are needed. However, with the increasing share of professional services on Airbnb, sociality is replaced with socio-technical mechanisms, such as a reviewing system and algorithmically managed search engine, related policies and guidelines (Christensen, 2022; Gössling et al., 2021).

Trust in the platform is more often based on its functions and ability to mediate interactions, while more emotional commitments are apparent in the interpersonal trust (Pelgander et al., 2022). This puts a greater emphasis on the platform's functions in the digital environment of consumption as a basis for trusting the overall setting in which consumers operate. The platform's role in developing trust exceeds providing access to a reviews-based reputation management system and requires additional functions and guidelines to use it.

This environment is often defined by terms of use, and codes of conduct that determine how the platform uses data and consumer-generated reviews, the possible ways to engage with different platform's elements and affordances.

More specifically, we echo the proposition by Gössling et al. (2021) that policies are a necessary component of the trust management system, as they define the platform's data use practices, quality indicators used to mark 'trustworthy' hosts, and designate allowed users' actions. On the other hand, community guidelines in the form of less formally communicated norms are important carriers of the platform's ideological ethos, which positions trust and belonging at the core of the platform's operations. In particular, the platform defines trust as corresponding to its ideas of safety, security, fairness, authenticity and reliability (Gössling et al., 2021). However, other than studies that show the effects of information asymmetry (Zamani et al., 2019), there is little tourism research about how the platform's policies and community standards affect consumption practices and how they correspond to each other. We analyse how these specific ideas of trust and trustworthiness are further formulated in the documents that provide more concrete suggestions and allowed or disallowed behaviours in specific situations.

Formalising consumer behaviour with discourse

To demonstrate how consumption practices are formalised by a platform's terms, community guidelines and policies, we first look at the literature that explores the strategic application of discursive elements common in these documents. Studies of platforms already show that digital platforms can significantly affect and mediate their users' behaviour on platforms (Bucher & Helmond, 2018). However, despite interest in content that regulates behaviour on platforms, this topic is largely overlooked in studies of consumer behaviour (Fitchett & Caruana, 2015). In our analysis we regard three types of content that formulates the platform's discourse – its legal terms of service, community policies and community guidelines. The content of these documents differs in their purpose, tone and principles that they convey. While guidelines provide an overall view of a platform's ideology, the community policies suggest how specific situations should be approached by users, and the Terms of Service signify the contractual roles of the user and the platform (Gillespie, 2019).

Community policies and guidelines sound similar but differ in their application and usual strategic forms. Both types relate to a platform's rules, which need to be followed for the platform organisation to be able to enforce behaviour (Gillespie, 2019). For this reason, it is important to consider how this content is used to communicate rules and norms. As noted by Gillespie (2019), policies such as Terms of Service matter more when decisions are disputed, while community guidelines are themselves a 'discursive performance'. Their primary purpose in relation to consumer is to reassure that the platform's environment is friendly towards them. This serves a few purposes. First, the platform leaves the content ambivalent, which allows more space for judging what behaviours are allowed or not (Gillespie, 2019). In addition, although ambivalent, the content supports trust in the platform itself, suggesting that it operates in the same framework of values as the consumers, addresses perceived risks and consumers' expectations towards service quality and cares for consumers. These elements are important antecedents for building trust in short-term relationships (Blomqvist & Cook, 2018; Cohen et al., 2014).

While guidelines are often written in a casual tone that expresses its aims and values, policies are more formal, as they need to provide reasonably clear guidelines for internal teams that make decisions about sanctioning content or users, while leaving enough space for possible changes over time (Gillespie, 2019). Policies develop over time, both reflecting external developments and providing a basis for normalising the platform's economic and cultural growth and interests. As such, Airbnb's Terms of Service and other policies reflect is views on professionalisation of service providers, use of consumers' data, privacy and safety (Light et al., 2018). By defining related platform's functions policies also create a formal basis for interacting with the platform, defining accepted or constricted behaviours. Together, community guidelines and policies of Airbnb platform create a discursive performance between the ethos of sharing and platform's interests in professionalisation of its service providers, reflected in formal descriptions of allowed behaviours and more casually conveyed norms.

The community guidelines articulate the platform's ethos (Gillespie, 2019), which in Airbnb's case is closely related to the ethos of the sharing economy, expressed in casually written but formal standards for trust. Policies establish these categories as a point of reference in specific situations and require users to follow them. Terms of service are more specific, as they attribute pre-defined roles to the platform and its users. As noted by Öberg (2021), although the roles of the users are similar to traditional business models, they become paradoxical when considering trust on platforms. Roles of guests and hosts overlap and are described in the terms as those of independent actors, which extend their roles beyond usual expectations and implicate trust. Roles are set by the platform's terms, making it important to study in the overall discourse and relate to the categories of trust building established in guidelines and policies.

As exchanges happen online, platforms assume an important role of mediators of public discourse, while at the same time also constituting such discourse themselves (Gillespie, 2019). Consumption depends in large part on discourse to create knowledge and meaning for products, and while consumption is experienced on an individual level, it is indeed constructed in discourse (Fitchett & Caruana, 2015). However, as Fitchett and Caruana (2015) and Skålen (2010) have noted, research into the connection between discourse and marketing and consumer behaviour is rather scarce. One of the reasons why such approach has been resisted in marketing and consumer research is that it questions some fundamental assumptions of the field, including the view of *consumer as chooser* (Fitchett & Caruana, 2015).

While platforms such as Airbnb state clearly that their role is that of a mere mediator and facilitator of exchange, they also contribute to the creation of public discourse (Gillespie, 2019). While rules of conduct are often only a background to social interaction on the platform and come to the fore only when conflict arises and rules need to be enforced, the ways in which such rules and guidelines are articulated are of great importance as they fulfil several functions for the platform: they legitimise the platform's role and work as moderators, they establish principles on what is or is not permitted and why, they state what is expected of users and what is considered acceptable behaviour (Gillespie, 2019).

Method: Discourse analysis

The study employs a constructivist discourse analysis to understand the formalisation of trust in Airbnb's policy documents and guidelines. Gillespie (2019) argues that platforms across the web seem to be; converging on a commonsense notion of what should be prohibited'. As a paradigmatic case of peer-to-peer accommodation platforms, Airbnb is the perfect ground to study what kind of discourse constitutes a 'trustworthy consumer' in the landscape of online platforms in the tourism industry. In fact, paradigmatic cases are appropriate for understanding and highlighting the general characteristics of the phenomenon studied (Flyvbjerg, 2006).

The constitutive orientation to discourse analysis proposes that discourse is constitutive of social reality and considers the situated nature of the text to understand how discourse is used in context to constitute certain kinds of subjects in certain kinds of relations to others (Fitchett & Caruana, 2015). While the technical orientation of DA focuses more on the microlevel of the individual text, and the political orientation concerns itself with the macrolevel of the evolution between marketing and society, constitutive or constructive DA focuses on the micro-mesolevel of how specific subjectivities emerge in marketing and consumer behaviour (Fitchett & Caruana, 2015; Skålén et al., 2006).

Therefore, we analyse documents in context, rather than focusing on specifics of language and pragmatics (Cameron, 2001). This means that the content of the documents is analysed as embedded in the interests of the platform organisation that produces them and known practices for trust building with socio-material mechanisms. For example, the analysis addresses what is shown in the foreground of the discussion, and what is left in the background (Van Dijk, 1996), and details are potentially left out for keeping the content ambivalent, discursive elements that externalise trust-related phenomena (Potter, 1996).

We approach discourse analysis by addressing the studied documents, their content, message and structure. The documents are of different types, which serve specific purposes and are meant for specific audiences, which affect the rhetoric considerations in their production (Potter, 1996). The documents' content thus varies in tone and ambivalence, while providing justifications specific to the addressed audiences (Gillespie, 2019). Further attention is given to the messages carried by the documents in addressing the principles they convey, the justifications for what is considered normal or abnormal in Airbnb's marketplace.

Sampling and data collection

A purposive sampling strategy led to the selection of a data corpus that fulfils specific theoretical criteria for analysis (Daniel, 2012). Since the analysis focuses on formalising trustworthy consumption behaviour, categories relevant for perceiving trustworthiness were identified as sources for sampling criteria. In this case, we have referred to earlier studies that identified Airbnb's community guidelines as a specific document that describes trustworthiness as a measure of safety, security, fairness, authenticity and reliability (Gössling et al., 2021), and started the data collection from there.

On the Airbnb website, documents have a label: 'rules', 'legal terms', 'how-to' or 'community policy'. We use these labels to identify the type of document. Such labels, however, are not always clear. At times pages that appear in a list of 'community policies' are labelled 'rules' (e.g., Our Community Standards), while some policies are legally binding and are therefore labelled 'legal terms' (e.g., Privacy policy). Therefore, in sampling the data we have based the distinction between types of documents on our own interpretation of the content as well as the label, thus identifying three types of documents based on their legal strength and type of content: (1) terms; (2) policies; (3) guidelines and standards.

The platform provides various other forms of content, such as marketing material, blog posts, a discussion forum for community, where users can discuss various issues, user-generated content and reviews. We have opted not to include this content in the analysis because it is often either specifically marketing-based, or user-generated, and reflects the views of platform's users, which do not contribute to the analysis of formalisation of behaviour. The collected guidelines, policy pages and terms were found on the Airbnb website. After reviewing 42 documents among legal terms, policies and guidelines, we focused the analysis on 15 key texts, a total of 65 pages in the data corpus. See Table 4.1 for more details.

The analysis focuses on several areas of questioning that follow the constructive approach to discourse analysis (Potter, 1996). Starting with some descriptive categories, the focus of the analysis is on the analytical categories

TABLE 4.1 Information about the main documents that relate to trust and the use of Airbnb platform

	Document	Type	Pages	Purpose?
1	Terms of Service	Legal Terms	24	Contract
2	Our Community Standards	Guidelines/ Rules	5	Standards for trustworthy behaviour
3	Nondiscrimination Policy	Community Policy	6	Principles for inclusion and respect
4	Airbnb's Off-Platform Policy	Community Policy	3	Defines allowed and prohibited actions not on Airbnb platform
5	Extenuating Circumstances Policy	Community Policy	3	Establishes grounds for overriding cancelation policies
6	Rebooking and Refund Policy	Community Policy	3	Sets a basis for helping customers in need of rebooking and refunds
7	Host and Guest Safety	Community Policy	2	Establishes norms for safety
8	Airbnb's Content Policy	Community Policy	3	Designates allowed and disallowed use of content
9	Airbnb's Reviews Policy	Community Policy	2	Designates norms for reviewing
10	Cancellation policies for your listing	Guidelines	3	Helps select appropriate cancelation policy
11	Find the cancellation policy for your stay	Guidelines	2	Collects information about cancellation policies
12	How search results work	Guidelines	3	Builds confidence in using Airbnb's search engine
13	Airbnb service fees	Guidelines	2	Explains the structure of fees on the platform
14	Responding to reviews	Guidelines	1	Explains how to respond to reviews
15	Host Privacy Standards	Legal terms	3	Sets expectations and prohibitions in use of personal data

of content and rhetoric, respectively pertaining to the 'what' and the 'how' of the content that is communicated through these documents. After a first round of in vivo coding, which yielded several initial codes, we then identified and focused on four analytical categories. Two were descriptive: (1) document type, audience and purpose; and (2) document structure, disposition, length and preambles. The other two were analytical: (3) content, including justification and message (expectations, appropriateness, principles, prohibitions); and (4) rhetoric, including tone, sentence structure and level of ambivalence.

Based on these analytical categories, four themes emerged, which guided the analysis in the last stage: foregrounding, backgrounding, principled justifications, roles. Through the analysis of foregrounding, we question what issues are emphasised, made important and explicitly stated. Similarly backgrounding focused on identifying issues and topics that are played down and minimised, albeit still present (Potter, 1996). We have identified principled justifications, which, according to Gillespie (2019), are statements often presented in the preamble of each document that justify the platform's mediating and rule-setting role through some principle that is at the basis of the platform's existence. Here the platform needs, on one hand, to proclaim its principles while at the same time establishing the need for rules and its own role as rule enforcer. Through the principled justification and other discursive devices, the different roles of the actors that interact on the platform, as well as the role of the platform itself in the exchange, are delineated.

Analysis

Our analysis starts with the document *Our Community Standards*, interpreting its contents as a benchmark for further analysis, as it contains general rules for behaviour and explicitly proposes specific standards for understanding trustworthiness. These standards guided us to further look into Community Policies searching for pages that would be directly linked to the categories related to trustworthiness, such as reviewing policy, non-discrimination policy, content policy. These and other documents show how consumers should treat each other, the platform and its content in specific situations that relate to building trust. Further on, as explained in the previous section, Airbnb's *Terms of Service* provide the most concrete content about roles, obligations and rights of the platform, consumers and providers; therefore, they were included in the analysis, looking for cues that relate to trust.

Foregrounding trustworthiness

The basis of Airbnb's community marketplace is trust. This is clearly expressed in the opening of main document that formulates rules of the marketplace:

> Creating a world where anyone can truly belong requires a foundation of trust grounded in consistent expectations of host and guest behaviour. We've established these Community Standards to help guide behaviour and codify the values that underpin our global community.
>
> *(Our Community Standards)*

The keywords in this statement are 'guiding' behaviour and 'codifying' values 'that underpin our global community'. Codification of values creates a guideline that helps express and share information about expectations that form the basis for trust in this marketplace (Williams & Baláž, 2021). This is done by defining five pillars for trustworthy behaviour: safety, security, fairness, authenticity and reliability. Each topic has a section, which is opened by identifying its relation to trust, for example: 'Whether you're opening your home as a host or experiencing a host's hospitality as a guest, you should trust that you will feel secure' (*security* section) or 'Your Airbnb experiences should be full of delightful moments and surprising adventures. Since our community is built on trust, authenticity is essential' (*authenticity* section). Trust is codified in these standards by repetition and explanation of related expectations.

The sections that outline expectations as 'pillars' for trust are set as a separate chapter of the standards and further explained. This projects an image of the marketplace as sharing a common set of values, which is an important antecedent of trust (Mayer et al., 1995). The same main pillars of trust are seen throughout community policies that codify expected actions in specific situations. However, as was noticed in the analysis, some pillars are more expressed than others. Notably safety and security are often foregrounded as the basis for policies that define use of content, reviewing, cancellations, off-platform behaviour and other instances.

For example, the Airbnb's *Off-Platform Policy*, which identifies allowed and prohibited interactions outside of the platform, is reasoned with the necessity to 'protect our community and business'. Safety and security are common concerns for platforms, as issues often occur due to dishonest behaviour or unethical data use (Wirtz et al., 2019), and by focusing attention on these pillars the platform suggests that it is both aware of and able to address these issues.

Safety and security are complemented by setting the standards for fairness and authenticity as treating 'everyone with respect in any interaction' (fairness) and 'shared expectations, honest interactions, and accurate details' (authenticity). They guide policy as well as legal terms for *Host Privacy Standards*:

> As a Host you will receive and use Guests' personal information to manage your reservations and deliver your Host Service. Please remember that you are responsible for complying with applicable privacy laws when you handle and process personal information.
>
> *(Host Privacy Standards)*

The platform acknowledges that hosts have power over their guests by having access to their personal data and can abuse it. The use of data is largely defined by external regulations such as the European Union's General Data Protection Regulation (GDPR), and legal terms mainly refer to such regulations. However, they frame this standard as the need to maintain the safety of guests, as well as the need to act fairly and provide authentic information about oneself.

The pillars for trust are described as factual constructions by using common rhetorical devices, such as the tone of speech and ambivalence of the document's content. For example, the Community Standards are written in what

Gillespie (2019) would describe as a casual, but firm, tone that allows for a broad range of language registers to be used in specific details, for example, describing prohibitions as something that should not be done, e.g., 'You should not share personal information to shame or blackmail others…'. In this case, even common expectations for responsible use of personal information are explained in a casual manner as something the guests or hosts can choose. The tone describes trust as a necessity and urges the audience to themselves act towards it (Potter, 1996). However, there is no space to misinterpret the coded standard for using personal information responsibly.

Similar statements are common throughout the guidelines, coupled with normative 'dos and don'ts' in describing desired or prohibited actions. They mostly refer to the coded expectations towards consumers and providers. As shown in the further analysis, the content that formulates expectations towards actions or decisions of Airbnb as the platform organisation is formulated in much more ambivalent language.

Backgrounding the value proposition

While trust is placed at the front of Airbnb's policies, urging consumers to act towards it, the documents maintain the platform's value proposition at the background. After all, the ideal Airbnb user, in fact, is not a trustworthy person but a trustworthy *consumer*. Thus, although backgrounded, the economic value of guest-host exchange is also at the basis of the discourse. Every aspect of the platform's functionality is related to its efficient and financially successful operation of the platform. The platform's revenue is directly connected to the revenue of its service providers:

> To help Airbnb run smoothly and to cover the cost of services like 24/7 customer support, we charge a service fee when a booking is confirmed.
>
> *(Airbnb Service Fees)*

While the service fees section specifically explains what fees are charged by Airbnb, in the most documents the platform's interest in commercial part of the exchanges is unmentioned. Even in the Service Fees section, the existence of fees is explained by Airbnb's value proposition of 'covering the costs of services'. Elsewhere the economic interests of the platform and hosts are mostly placed in the background of the more emphasised notions of trust and community.

Another indirect mention of the platform's value proposition appears when referencing necessity to maintain high ratings in *Our Community Standards*. The document ends with an instruction that hosts 'should not have persistently and pervasively low ratings'. This is framed as part of the guidelines on 'being responsive'; however, the notion of maintaining high ratings does not directly relate to responsiveness. Instead, it appears to relate to upholding a standard at an unindicated level of quantified ratings. Being responsive is indicated as one of the pillars for being trustworthy, which is measured by the platform's review and rating system. Focusing attention on responsiveness as a part of overall ratings allows for leaving the overall interest in maintaining a highly rated supply of hosts in the background of trust. Caution not to have low ratings indicates an expectation for performance that does not need to be elaborated upon, when tied into a discourse of trust.

Both the Airbnb platform and its users gain value from a growing network of users (Reinhold & Dolnicar, 2017). 'Growth' is therefore an important aspect of the platform's business, and a clear topic that reappears within its policies, although is rarely emphasised. The Nondiscrimination policy is explained as necessary 'As the Airbnb community grows…'. Growth is also aligned with personal growth of community members: 'Honest reviews help Hosts and guests to grow and find the right fit in the future' (Responding to reviews). The metaphor of 'Growth' is interesting, as, although the term remains generic, it is also emphatic. Growth emphasises that trust is also aligned with the platform's basic economic function of matching guests and hosts that 'find the right fit' (How search results work).

The platform's policies thus emphasise trust and urge consumers and providers to act towards maintaining trust proposed by presented standards. While trust constitutes the fore of the discourse, its value proposition of providing services that match consumers and providers and collecting fees from a growing number of exchanges remains at the background of both the platform's functionality and the documents that define it.

Principled justifications

The platform acts both as a service provider and an intermediator of guest–host relations, and the documents we analysed use discourse to justify the role of the platform as mediator of trust, where a balance needs to be stricken between trust in the platform itself as a service provider, and interpersonal trust between hosts and guests. The justification thus contributes to the discourse of trustworthy consumer *on the Airbnb platform*. Such discursive performance is mostly performed in the first paragraphs of the document, the preamble. While in some of the documents the preamble is explanatory and to the point, for example in the Airbnb's Off-Platform Policy, 'In order to protect our community and business, the following behaviours are prohibited'; in many documents, the preamble serves a specific purpose in the discursive effort of formalising trust. In several of Airbnb's policies and guidelines these preambles set a solemn tone by using emphatic language: using many words to say one thing; repeating concepts; paraphrasing; using non-bureaucratic terms (e.g., harmony, empathy). At times the tone is instead patronising, emphasising the responsibility of the user and the platform's ambiguous role as neutral but at the same condescending or superior.

The justification is often explicitly juxtaposed to the principle in each sentence, for example: (a) 'While we....': principle; (b) '...we do not...': limits to the scope of the platform's role; (c) '...instead, we...': how the platform sets and enforces rules. For example, in the Nondiscrimination policy:

> While we do not believe that one company can mandate harmony among all people, we do believe that the Airbnb community can promote empathy and understanding across all cultures. (...) To that end, all of us, Airbnb employees, hosts and guests alike, agree to read and act in accordance with the following policy.
>
> *(emphasis added)*

The most explicit and complex principled justification is in the preamble of Our Community Standards. After the emphatic opening seen above, where the document states the platform's aim of 'creating a world where anyone can truly belong' and its effort to do so through the Community standards, the second paragraph states:

> To help ensure safe stays, experiences, and interactions – safety, security, fairness, authenticity, and reliability remain central pillars in our efforts to ensure safety and foster belonging. We're always working to make sure they're upheld and enforced.
>
> *(Our Community Standards)*

Although the preamble consists of only two short paragraphs, it does important discursive work and uses a number of rhetorical devices. Each paragraph starts with an action statement, which describes Airbnb's role as community builder and helper in the forefront. This is done with the two locutions: 'Creating a world' and 'To help ensure'. The two paragraphs continue, in a specular way, indicating the object of these actions: 'anyone' and 'stays, experiences and interactions'. Each paragraph also includes two distinct feelings that are at the basis of the experience: belonging and safety. The emphasis added to these statements by using repetitions, paraphrases and synonyms draws attention to these ideological elements in building trust. 'Creating a world where anyone can truly belong' is not attributed just to Airbnb; instead, as Celata et al. (2017) suggest, this formulation includes the consumers and the providers into a social endeavour to move towards such a world.

The justifications lay grounds for the discursive construction of trust and the trustworthy Airbnb consumer, actual or prospective – in *Our Community Standards* generically indicated as 'anyone'. In fact, after the emphatic statement about belonging and world building, the text goes on to clarify that these require a foundation of trust. It further specifies this trust as being grounded in the behaviour of hosts and guests, which should conform to consistent expectations (Our Community Standards).

Another rhetorical device is the sentence structure, where the statement of purpose or principle is followed by the justification of Airbnb's role as rule maker ('we've established these...') and enforcer: 'We're always working to make sure they're upheld and enforced' (Our Community Standards). In both paragraphs, the pronoun 'we' is used to indicate only the platform as an entity. While prescriptions on how the 'you' should act and feel do follow in the rest of the text, they are absent from the preamble, where the focus is on the principles (belonging, safety, trust) and the platform's role in ensuring them.

In this effort to formalise trustworthy behaviour, the document foregrounds the five pillars: safety, security, fairness, authenticity, reliability. Moreover, unlike most other texts that follow the principled justification in the preamble with more factual text, Our Community Standards continues by including another principled justification for each section of the document, that is, the five pillars of trust. For example, regarding safety, the identified principle is that using Airbnb should be an 'adventure'; however, the justification for the rules is that it needs to be safe.

The non-discrimination policy is another document that draws the reader into a discursive performance of the ideological basis for trust. As it stands at the time of writing, the policy has over a page-long preamble that offers an exceptionally detailed justification for its existence. Non-discrimination is a direct plea to avoid specific discriminatory behaviours. It proposes the platform's position on contested matters that hosts may disagree with (Zhu, 2021), but is necessary to reassure consumers about platform's stance on the issue of discrimination. This is not surprising, given the nature of the document which does not deal with the terms of the exchange that happens on the platform, but more generally with how people treat each other, within and beyond the scope of Airbnb hosting, with morals and ethical principles, with a person's values.

As noted by Edelman et al. (2017) and Ert and Fleischer (2019), Airbnb's trust management system has previously received several upgrades to reduce discrimination on the basis of race, gender and other common characteristics. The non-discrimination policy explicitly states that 'the public, our community, and we ourselves, expect no less than this', showing that the policy is a response to such behaviours and therefore fulfils an important role in the strategic communication of the platform's values, directed not only at the users but at the general public as well.

The preambles, therefore, commonly justify the necessity to have terms, rules and guidelines on the basis of some principles, which in turn are often related to the foregrounded five pillars of trustworthiness. These 'principled justifications' (Gillespie, 2019), however, often show some ambiguity on the role of the platform as rule maker and enforcer versus its role as mere platform for peer exchange.

Defining Roles of Consumers, Providers and the Platform

Trust is often grounded in having clear expectations towards others based on their distinct roles (Kramer, 1999). For example, expectations towards hosts can be set around their proven ability to provide the core service of hosting, indications of friendliness (Sparks & Browning, 2011). Roles on Airbnb are explicitly categorised between guests and hosts as platform's *Members*, and the platform, indicating specific expectations marked for each of them. This is done in both legal terms and guidelines, for example, with the Terms of Service specifically defining these roles:

> The Airbnb Platform offers an online venue that enables users ('Members') to publish, offer, search for, and book services. Members who publish and offer services are 'Hosts' and Members who search for, book, or use services are 'Guests.'
>
> *(Terms of Service)*

Terms of Service are the contract that Airbnb users agree to follow when using the platform, and the terms indicated here can be considered as a part of the legal discourse of Airbnb. They clearly indicate that guests and hosts are differentiated between those that publish their services and those that search for them, while the platform takes the role of enabling this behaviour. The same idea is reproduced further with more subtle discursive elements:

> As a Host, Airbnb offers you the right to use the Airbnb Platform to share your Accommodation, Experience, or other Host Service with our vibrant community of Guests - and earn money doing it. It's easy to create a Listing and you are in control of how you host - set your price, availability, and rules for each Listing.
>
> *(Terms of Service)*

This passage clarifies the roles, while also indicating the general expectations related to them. Hosts share their accommodation, experience or other service, with the community of guests, described as vibrant. Meanwhile Airbnb offers the right to use the platform and access 'our' community. Access to the community is presented as the platform's service, also indicating that it is commonly shared 'our' community, not clarifying if that refers to Airbnb alone, or includes the guests and the hosts, and who is responsible for upkeeping expected behaviour within that community.

Within these terms, Airbnb takes the role of an organiser of exchanges – it provides access to the community and the platform with its underlying information and communication technology infrastructure. This also establishes a hierarchy, in which Airbnb is expected to establish and enforce rules. This is reinforced by Airbnb's role in handling payments and refunds. Consider the following extract:

> This Rebooking and Refund Policy explains how we will assist with rebooking a reservation and how we handle refunds when a Host cancels a reservation or another Travel Issue disrupts a stay. <...>
>
> If we determine that a Travel Issue has disrupted the stay, we will provide a full or partial refund and, depending on the circumstances, may assist the guest with finding comparable or better accommodations
>
> *(Rebooking and Refund Policy)*

The aforementioned policy describes Airbnb's role in refunding stays cancelled not in accordance with its cancellations policies, in which case the guests can receive a refund or have a different accommodation rebooked by Airbnb. The policy is phrased in a way that requires the guests in question to ask for its help and reassures them that help will be provided.

The policy is positively framed, by making a claim that the platform is trustworthy to fulfil its role (Sparks & Browning, 2011). However, the text is also ambivalent about how this is to be carried out. Specific information about the speed of refund, type of rebooked accommodation or other more technical details is not given. As Gillespie (2019) suggests, policies of platforms are phrased this way to leave space for their employees to themselves navigate the platform's commitments in deciding when they should act or not.

The ambivalence around the definition of role of the platform and the strictness around the roles of consumers indicate that roles in this environment are constituted by the platform's discourse. Roles can change over time, as the platform needs to react to different external regulations. The role of the platform itself is defined with some ambivalence that allows freedom in deciding how to approach certain situations. Öberg (2021) notes that roles of consumers and providers in the platform economy are generally the same as in usual commercial exchanges; however, certain differences exist. The roles of consumers and providers on platforms can overlap, which appears to be the case when constituting expectations for the basis of trust, and it is important for the platform to leave itself space to adjust their definitions when necessary.

Concluding discussion

In this chapter we have approached the question of how a platform formalises trustworthiness with its discursive work in documents such as legal terms, policies and guidelines. Perhaps the most significant finding is the way in which policies and terms balance the interests of the platform organisation and expectations towards consumers. Airbnb's policies foreground trust and sharing as part of its rhetoric, prompting consumers to behave in a way that upholds this rhetoric, but also tie it closely with the way Airbnb platform creates and captures value. This discourse maintains a balance between trust in the platform as a service provider and interpersonal trust between hosts and guests. This balance is maintained by the tone and content of policies, providing both clear and normative guidelines for what is allowed and the roles of consumers and the platform, but also maintaining ambivalence when necessary.

The analysis of Airbnb's policy documents reveals a nuanced discursive strategy employed by the platform organisation to justify its dual role as a service provider and mediator of trust. The expectations towards behaviour are grounded with explicit principled justifications and serve as a basis for building common values and integrity, which are essential elements of trust. As we show, the principles that underpin trust are formalised by maintaining a delicate balance between trust in the platform and interpersonal trust between hosts and guests. The use of emphatic language, repetition and rhetorical devices in these preambles highlights the platform's stated commitment to creating a world of belonging and safety, while also justifying its role as a rule maker and enforcer.

Finally, we show that the discourse does extensive work in defining the roles of consumers, providers and the platform. Öberg (2021) has shown that trust in platform-based marketplaces is tied to these roles in the proclaimed community. Yet, as this discourse analysis shows, although role distinctions can appear straightforward, they are flexible, leaving some navigation space for making decisions in contested situations. It appears that roles are not static but shaped by the evolving discourse of the platform and influenced by external factors.

Earlier research about trust on P2P platforms mostly analyse platforms' functionalities, such as their online-reviews based systems for creating and managing reputation between consumers (Baute-Díaz et al., 2019; Zervas et al., 2021). In these studies, trustworthiness is viewed as a result of basing expectations of reputation generated through online reviews (Zloteanu et al., 2018). Our analysis shows that in addition to such functionalities, platforms maintain a narrative of trust, where trustworthiness is co-constituted by discourse as a dynamic phenomenon. Thus, our contribution to the study of consumer behaviour on platforms lies primarily in relating trust in consumer behaviour to discourse. Earlier studies suggest that consumers rarely read a platform's policy documents (Zamani et al., 2019); however, in this study we show that these documents perform significant discursive work in constituting the notion of trust. They craft an ideal *trustworthy consumer* by providing formal indicators of what consumer behaviours make a user worthy of trust, while at the same time balancing the different roles that users and platforms respectively hold in the exchange. This discourse does not need to be constantly read but remains the main source of formalised knowledge about the expected behaviours and roles.

References

Airbnb. (2023). *Our Community Standards*. https://Www.Airbnb.Com/Help/Article/ 3328

Baute-Díaz, N., Gutiérrez-Taño, D., & Díaz-Armas, R. (2019). Interaction and reputation in airbnb: An exploratory analysis. *International Journal of Culture, Tourism, and Hospitality Research*, *13*(4), 370–383. https://doi.org/10.1108/IJCTHR-10-2018-0149

Blomqvist, K., & Cook, K. S. (2018). Swift Trust: State-of-the-Art and Future Research Directions. In Rosalind H. Searle, Ann-Marie I. Nienaber, & Sim B. Sitkin (Eds.), *The Routledge Companion to trust*. Routledge.

Bosma, J. R. (2022). Platformed professionalization: Labor, assets, and earning a livelihood through Airbnb. *Environment and Planning A*, *54*(4), 595–610. https://doi.org/ 10.1177/0308518X211063492

Bucher, T., & Helmond, A. (2018). The Affordances of Social Media Platforms. In J. Burgess, A. Marwick, & T. Poell (Eds.), *The SAGE handbook of social media* (pp. 254–278). SAGE Publications Ltd. https://doi.org/10.4135/9781473984066

Cameron, D. (2001). *Working with spoken discourse*. SAGE.

Celata, F., Hendrickson, C. Y., & Sanna, V. S. (2017). The sharing economy as community marketplace? Trust, reciprocity and belonging in peer-to-peer accommodation platforms. *Cambridge Journal of Regions, Economy and Society*, *10*(2), 349–363. https://doi.org/10.1093/cjres/rsw044

Christensen, M. D. (2022). Doing digital discipline: how Airbnb hosts engage with the digital platform. *Mobilities*, 1–16. https://doi.org/10.1080/17450101.2022. 2060756

Cohen, S. A., Prayag, G., & Moital, M. (2014). Consumer behaviour in tourism: Concepts, influences and opportunities. In *Current issues in tourism* (Vol. 17, Issue 10, pp. 872–909). Routledge. https://doi.org/10.1080/13683500.2013.850064

Daniel, J. (2012). Sampling essentials: Practical guidelines for making sampling choices. SAGE. https://doi.org/10.4135/9781452272047

Edelman, B., Luca, M., & Svirsky, D. (2017). Racial discrimination in the sharing economy: Evidence from a field experiment. *American Economic Journal: Applied Economics*, *9*(2), 1–22. https://doi.org/10.1257/app.20160213

Ert, E., & Fleischer, A. (2019). The evolution of trust in Airbnb: A case of home rental. *Annals of Tourism Research*, *75*(January), 279–287. https://doi.org/10.1016/j.annals. 2019.01.004

Ert, E., Fleischer, A., & Magen, N. (2016). Trust and reputation in the sharing economy: The role of personal photos in Airbnb. *Tourism Management*, *55*, 62–73. https://doi.org/10.1016/j.tourman.2016.01.013

Fam, K. S., Foscht, T., & Collins, R. D. (2004). Trust and the online relationship—An exploratory study from New Zealand. *Tourism Management*, *25*(2), 195–207. https:// doi.org/10.1016/S0261-5177(03)00084-0

Fitchett, J., & Caruana, R. (2015). Exploring the role of discourse in marketing and consumer research. *Journal of Consumer Behaviour*, 14(1), 1–12. https://doi. org/10.1002/cb.1497

Flyvbjerg, B. (2006). Five misunderstandings about case-study research. *Qualitative inquiry*, 12(2), 219–245. https://doi.org/10.1177/1077800405284363

Fraanje, W., & Spaargaren, G. (2019). What future for collaborative consumption? A practice theoretical account. *Journal of Cleaner Production, 208*, 499–508. https://doi.org/10.1016/j.jclepro.2018.09.197

Gil, J., & Sequera, J. (2020). The professionalization of Airbnb in Madrid: Far from a collaborative economy. *Current Issues in Tourism*, 1–20. https://doi.org/10.1080/13683500.2020.1757628

Gillespie, T. (2019). *Custodians of the Internet*. Yale University Press. https://doi.org/10.12987/9780300235029

Gössling, S., Larson, M., & Pumputis, A. (2021). Mutual surveillance on Airbnb. *Annals of Tourism Research, 91*(November 2021), 103314. https://doi.org/10.1016/j.annals.2021.103314

Huang, D., Coghlan, A., & Jin, X. (2020). Understanding the drivers of Airbnb discontinuance. *Annals of Tourism Research, 80*(January 2020), 102798. https://doi.org/10.1016/j.annals.2019.102798

Korsgaard, M. A. (2018). Reciprocal Trust. In R. H. Searle, A.-M. I. Nienaber, & S. B. Sitkin (Eds.), *The Routledge Companion to Trust* (pp. 14–28). Routledge. https://doi.org/10.4324/9781315745572-3

Kramer, R. M. (1999). Trust and distrust in organizations: emerging perspectives, enduring questions. *Annual Review of Psychology, 50*(1), 569–598. https://doi.org/10.1146/annurev.psych.50.1.569

Light, B., Burgess, J., & Duguay, S. (2018). The walkthrough method: An approach to the study of apps. *New Media and Society, 20*(3), 881–900. https://doi.org/10.1177/1461444816675438

Mayer, R. C., Davis, J. H., & Schoorman, F. D. (1995). An Integrative Model of Organizational Trust. *Academy of Management Review, 20*(3), 709–734. https://doi.org/10.2307/258792

Moorman, C., Deshpandé, R., & Zaltman, G. (1993). Factors affecting trust in market research relationships. *Journal of Marketing, 57*(1), 81–101.

Morgan, R. M., & Hunt, S. D. (1994). The commitment-trust theory of relationship marketing. *Journal of Marketing, 58*(July), 20–38.

Öberg, C. (2021). Disruptive and paradoxical roles in the sharing economies. *International Journal of Innovation Management, 25*(4). https://doi.org/10.1142/S1363919621500456

Pelgander, L., Öberg, C., & Barkenäs, L. (2022). Trust and the sharing economy. *Digital Business, 2*(2). https://doi.org/10.1016/j.digbus.2022.100048

Perren, R., & Kozinets, R. V. (2018). Lateral exchange markets: How social platforms operate in a networked economy. *Journal of Marketing, 82*(1), 20–36. https://doi.org/10.1509/jm.14.0250

Potter, J. (1996). *Representing reality. Discourse, Rhetoric and Social Construction*. Sage.

Reinhold, S., & Dolnicar, S. (2017). Airbnb's business model. In S. Dolnicar (Ed.), *Peer-to-peer accommodation networks: Pushing the boundaries* (pp. 27–38). Goodfellow Publishers. https://doi.org/10.23912/9781911396512–3601

Schor, J. B., & Vallas, S. P. (2021). The sharing economy: Rhetoric and reality. *Annual Review of Sociology, 47*, 369–389. https://doi.org/10.1146/annurev-soc-082620

Skålén, P. (2010). A discourse analytical approach to qualitative marketing research. *Qualitative Market Research: An International Journal*, 13(2), 103–109. https://doi.org/10.1108/13522751011032566

Skålén, P., Fellesson, M., & Fougère, M. (2006). The governmentality of marketing discourse. *Scandinavian Journal of Management*, *22*(4), 275–291. https://doi.org/10.1016/j.scaman.2006.07.001

Sparks, B. A., & Browning, V. (2011). The impact of online reviews on hotel booking intentions and perception of trust. *Tourism Management*, *32*(6), 1310–1323. https://doi.org/10.1016/j.tourman.2010.12.011

Teubner, T., & Flath, C. M. (2019). Privacy in the sharing economy. *Journal of the Association for Information Systems*, *20*(3), 213–242. https://doi.org/10.17705/1jais.00534

Van Dijk, T. (1996). Discourse, power and access. In C. R. Caldas-Coulthard & M. Coulthard (Eds.), *Texts and practices: Readings in critical discourse analysis*. Routledge.

Williams, A. M., & Baláž, V. (2015). Tourism risk and uncertainty: Theoretical reflections. *Journal of Travel Research*, *54*(3), 271–287. https://doi.org/10.1177/0047287514523334

Williams, A. M., & Baláž, V. (2021). Tourism and trust: Theoretical reflections. *Journal of Travel Research*, *60*(8), 1619–1634. https://doi.org/10.1177/0047287520961177

Wirtz, J., So, K. K. F., Mody, M. A., Liu, S. Q., & Chun, H. E. H. (2019). Platforms in the peer-to-peer sharing economy. *Journal of Service Management*, *30*(4), 452–483. https://doi.org/10.1108/JOSM-11-2018-0369

Zamani, E. D., Choudrie, J., Katechos, G., & Yin, Y. (2019). Trust in the sharing economy: the AirBnB case. *Industrial Management and Data Systems*, *119*(9), 1947–1968. https://doi.org/10.1108/IMDS-04-2019-0207

Zervas, G., Proserpio, D., & Byers, J. W. (2021). A first look at online reputation on Airbnb, where every stay is above average. *Marketing Letters*, 32(1), 1–16. https://doi.org/10.1007/s11002-020-09546-4

Zhu, H. (2021). Why a non-discrimination policy upset Airbnb hosts? *Annals of Tourism Research*, *87*, 102984. https://doi.org/10.1016/j.annals.2020.102984

Zloteanu, M., Harvey, N., Tuckett, D., & Livan, G. (2018). Digital identity: The effect of trust and reputation information on user judgement in the sharing economy. *PLoS ONE*, *13*(12), 1–19. https://doi.org/10.1371/journal.pone.0209071

5

EXAMINING THE IMPACT OF TRIP EXPERIENCES ON OVERALL SATISFACTION AMONG INDIAN TOURISTS IN NEW DELHI'S HERITAGE SITES AND HOTELS

Monisha Juneja and Tahir Sufi

Introduction

In many countries, tourism is a prime source of revenue that generates employment. A substantial market share of the international economy comes directly through tourism (Santa-Cruz and López-Guzmán, 2017). Investment in the tourism sector benefits from both inbound and domestic tourist activities and is crucial to a country's economic growth. Therefore, tourism is served by employing good management practices, eliciting positive publicity and revisiting intentions (Fang et al., 2020).

Tour operators play a crucial role in promoting tourism services as they are the driving force behind successful vacations. Their expertise and resources enable them to provide quality products and services that meet tourists' expectations, ultimately leading to satisfaction (Taheri et al., 2019). Tour operators also act as a bridge between tourists and service providers, offering a multidimensional experience that extends beyond trip itineraries and hotel stays (Chen & Rahman, 2018; Marin-Pantelescu et al., 2019).

Despite the various studies conducted on the factors that affect tourist satisfaction, there is limited consensus on the specific aspects of tourist behaviour and motivations that impact this phenomenon (Bayih & Singh, 2020; Osman et al., 2020). However, there is a clear understanding that tour operators, guides and agencies play a crucial role in ensuring a positive experience for tourists (Marin-Pantelescu et al., 2019). Sangpikul (2018) explored the impact of destination attractiveness on tourist satisfaction, while other studies have investigated the influence of tourist engagement, experiences and interactions on overall satisfaction. Alrawadieh et al. (2019a) examined the connection between tourist engagement and satisfaction. Similarly, Cajiao et al. (2022) discovered

DOI: 10.4324/9781032637778-6

a favourable link between travel experience and motivation. Wu and Li (2017) looked into how the sociodemographic profile of tourists affected their satisfaction. Domínguez-Quintero et al. (2020) also identified the authenticity of heritage sites as a factor influencing visitor satisfaction. Trip experience as a mediating variable of tourist satisfaction has been explored in various studies (Alrawadieh et al., 2019b; Cajiao et al., 2022). By contrast, Domínguez-Quintero et al. (2020) found no relationship between trip experience and satisfaction.

This study aimed to investigate the relationship between tourist satisfaction and trip experience considering its multidimensional nature in the Indian context. While prior research has found a positive relationship between tourist satisfaction and factors such as demographic profiles, site authenticity and destination loyalty, the connection between trip experience and satisfaction has been less clear. Some studies have found no relationship between the two, whereas others have suggested that trip experience is related to motivation. However, this study aimed to explore the mediating role of trip experience in the relationship between satisfaction and other factors. To accomplish this, the study examined the attributes of trip experience and overall satisfaction separately, and then explored the relationship between the two for all tourists who booked their Delhi trip through a travel agent. The results of this study could be useful for tourism intermediaries to focus on improving their trip experience, which in turn could affect the services provided by tour operators.

Literature review

Dimensions of tourist satisfaction

Tourist satisfaction is a positive feeling that arises from an individual's collective assessment of their experience, which leads to a sense of well-being (Oliver, 1993; Lee et al., 2012; Chen-Yi, 2016; Zhou et al., 2023). It serves as an indicator of effective destination management, as evidenced by increased popularity and repeat visit intentions (Fang et al., 2020; Loncaric et al. 2019). When tourists are satisfied, they are more likely to recommend the destination to others and share it on social media (Hollebeek and Rather, 2019; Sthapit et al., 2019). Moreover, pricing tourist products and services affects tourist satisfaction (Su et al. 2022).

According to previous research, satisfied tourists are likely to promote their destinations by attracting other potential visitors. This promotion can occur through word-of-mouth or other means, and can result in increased tourism revenue for the destination. On the other hand, providing high-quality tourism services can lead to increased customer satisfaction, but can also result in lost revenue for service providers. Additionally, memorable and long-lasting tourist experiences can have a significant impact on destination revisits, and satisfied tourists can act as brand ambassadors to the destination, attracting more visitors. This is supported by studies that show that satisfied tourists are more likely to return to a destination and recommend it to others (Barnes et al., 2020; Alexander et al., 2020).

The concept of customer loyalty as a marketing strategy to expand the customer base is frequently discussed in the tourism literature. Additionally, satisfied travellers can contribute to increasing the customer base by spreading positive word-of-mouth and bringing in their contacts, friends and relatives. The term 'destination loyalty' is commonly used in tourist literature to describe this phenomenon (Valverde-Roda et al., 2022). Ultimately, customer recommendations and the quality of the tourism product being resold (visit) play a significant role in generating revenue for the travel and tourism industry (Sangpikul, 2018).

Digital technology and tourist satisfaction

Tourism has experienced numerous benefits from digital technology, particularly destination accessibility (Gračan et al., 2021). Technological advancements have revolutionised the way tourists experience destinations (Dieck & Jung, 2018). Tourism service providers have been able to conduct their businesses with greater ease and reach a larger potential tourist market (Ray, 2022). As Garau (2017) noted, UNESCO and the WTO have urged service providers to leverage digital technology to meet tourist demands. The role of technology in travel planning has never been more significant, with tourists harnessing the potential to plan their journeys, access entertainment and share their experiences on social media. Technological advancements have a significant impact on tourist satisfaction and destination promotion (Mrsic et al., 2020) by creating long-lasting memories that can be shared in real time (Pai et al., 2020). Gračan, Zadel and Pavlović (2021) conducted research on how modern technology, such as intelligent maps, transport booking, and experience sharing, can contribute to enhancing tourist satisfaction.

The Internet pushed the global online travel agent (OTA) market to US$744.7 billion in 2019, driven by significant OTAs, including TripAdvisor, Expedia and Yelp, India, which has the second-largest Internet users globally (Garcia, 2023). Its impact can be seen in growing Internet use in the travel segment (Basuroy, 2022). Travel was once private and based on trust; tourists shared their experiences with friends and acquaintances. However, the Internet has converted these experiences into a global database. Travellers from anywhere in the world can access the database. Furthermore, the process allows users to record and share their experiences online (Selim et al., 2022). Posting online content about one's travel was done for fun and excitement as concluded by Oliveira et al. (2020). Such posts motivate other tourists in travel-related decision-making.

Tourist satisfaction: Role of tour operators

Tour operators play a significant role in shaping the image of a destination by disseminating positive information about the sites to travellers. They offer services that they purchase in large quantities and then package and sell as organised tours (Marin-Pantelescu et al., 2019). Tour operators function as intermediaries connecting buyers and sellers. They engage in extensive interactions with tourists, thereby expanding their knowledge of their destination. Tourists benefit from the convenience and planning offered by these tours, which saves time and money (Page & Connell, 2020). Collaboration between tour operators and various stakeholders, including hotels, bus operators, food outlets, retail outlets, sites and guides, ensures that tourists receive the best possible services (Devaraja & Deepak, 2018). Tour operators and other essential tourism stakeholders play critical roles in ensuring tourist satisfaction. They gather feedback from tourists to improve their trip planning and effectively manage their destinations. Continuous improvements in travel operations can enhance tourist experience and overall satisfaction (Cetin, 2020; Zhang et al. 2022).

The job of a tour guide or tour leader can significantly affect a tourist's experience. They often interact closely with tourists and their performance, as well as the satisfaction of tourists, depending on group tours. The evaluation of the entire tour, including the itinerary, hotel stay and meals, depends on the tour operator's assessment of the guide's explanation and ability to generate interest. Ultimately, the personal experience of the tour is also a factor in determining its success. A study by Cetin and Yarcan (2017) identified the specific qualities that a tour guide should possess to contribute to the success of tour packages. These professionals differentiate themselves from their competitors, and play a crucial role in selling future trips. Their professional behaviour is critical for exceeding tourist expectations and ensuring the success of the tour.

Strong collaboration between tour guides and operators helps create fantastic tourist experiences and enhances the destination's image. Therefore, the tour operators have been called 'cultural mediators' (Weiler & Yu, 2007; Huang et al., 2010; Weiler & Black, 2021). Tourists want to experience authentic local lifestyles, customs and culture with the help of tour leaders and guides (Syakier & Hanafiah, 2022), which can only be translated into tourist surroundings.

Tourist satisfaction: Local culture, food, and hospitality

Tourists visiting heritage sites and experiencing history are drawn to cultural heritage sites, festivals and events (Io, 2019). The tourist heritage experience must include performance arts such as live shows of dance, music and plays that portray heritage (Zhang et al., 2020). Similarly, shopping is a popular tourist activity, and accessibility to these stores is an important component of the tourism experience (Mawufemor et al., 2019); tour operators often include shopping on their itineraries (Fairhurst et al., 2007; Kong & Chang 2016). Furthermore, it contributes to the economic uplift of local communities and promotes local products (Kaur et al., 2016).

Buying souvenirs, keepsakes and mementos is associated with tourist travel, making the trip a tangible experience that can be gifted to others or extended to one's memory (Wilkins, 2011; Chang et al., 2022). Tourist experience is not limited to products but includes the services delivered by vendors (Mawufemor et al., 2019). Hotel stays are important in framing tourist experiences (Elghani, 2012; Lee & Chhabra, 2015; Pourabedin et al., 2022). A hotel stay creates a lasting image in the guest's mind, as many services are provided during the stay (Ugwuanyi et al., 2021). Fuentes-Moraleda et al. (2020) recommended that the importance of food in travel experiences should be noted.

Hotels offer distinctive heritage experiences in terms of accommodation, amenities and dining options. They contribute significantly to shaping brand image by disseminating favourable destination-related information to tourists. This, in turn, leads to a deeper comprehension of the tourist experience (Elghani 2012). A research study by Yameen (2013) found that tourists were content with their experiences. The three key elements that determine satisfaction are (i) the people involved, (ii) the physical environment in which the service is provided and (iii) the service-delivery process. Hotels are vital in defining the tourist experience (Lee and Chhabra, 2015; Pourabedin et al., 2022).

Gastronomic interests and culinary encounters play a crucial role in shaping the perceived value of historical and cultural destinations, which is essential for tourist satisfaction. According to Mora et al. (2021), gastronomy serves as a fundamental pillar of tourism, affecting satisfaction with and loyalty to the destination. Both tangible and intangible cultural experiences are valued highly by tourists. Research studies categorise intangible cultural heritage into two categories: extrinsic and intrinsic. Live performances possess extrinsic attributes (Lee et al., 2016; Park & Petrick, 2016), whereas performing arts and artistic craftsmanship are considered intrinsic attributes (Bergadaa & Lorey, 2015). These experiences enhance tourists' experiences and enjoyment, contributing to their popularity (Park & Petrick, 2016; Zhang et al., 2020).

Trip experiences and tourist satisfaction

Murphy et al. (2000) comprised both physical and emotional dimensions. Swarbrooke (2001) asserts that various dimensions of the tourist experience are formed based on travel. The tangible aspects of the trip are the physical dimensions, which include welcoming tourists, transportation and accommodation (Cetin & Bilgihan, 2015; Liao & Chuang, 2020), places of interest (Ghose &

Johann, 2019), tour leader role (Yen et al., 2018; Teng & Tsai, 2020), tour guides (Huang et al., 2010; Cetin & Yarcan, 2017; Weiler & Black, 2021) and trip excitement (Park & Petrick, 2016; Io, 2019; Zhang et al., 2020). Tourists build on each of these experiences, resulting in their actual experiences.

The intangible or emotional dimensions of the tourist experience include feeling relaxed, engaging in adventure activities and visiting other attractions (Arnould and Price 1993; Yoon and Uysal 2005; Kim et al. 2012). The study by Carbone and Haeckel (1994) divides such dimensions into two broad groups, including 'Mechanic' and 'Humanic'. While the 'Mechanic' dimension categorises the physical environment, the 'Humanic' dimension deals with social interaction.

The debate on the different types of tourism experiences remains despite numerous studies on the characteristics of these experiences (Cetin al., 2012; Cetin al., 2015; Konstantakis et al., 2020). A comprehensive tourist experience includes various elements such as destination infrastructure, accessibility, accommodation and points of interest. However, existing research has primarily focused on trip duration and mode of travel to assess overall satisfaction (Morris & Guerra, 2015; De Vos, 2017; Alrawadieh et al., 2019; Ghose & Johann, 2019). While extensive research has been conducted on tourist satisfaction and trip experience, a gap remains in understanding their interconnectivity. This study has provided a framework to bridge this gap by examining the relationship between trip experience and overall tourist satisfaction for tourists visiting Delhi, who were booked through tour operators (Figure 5.1). Based on previous research, the following hypothesis is formulated:

H_o: There is no significant relationship between the trip experience and overall tourist satisfaction

H_a: There is a significant relationship between the trip experience and overall tourist satisfaction.

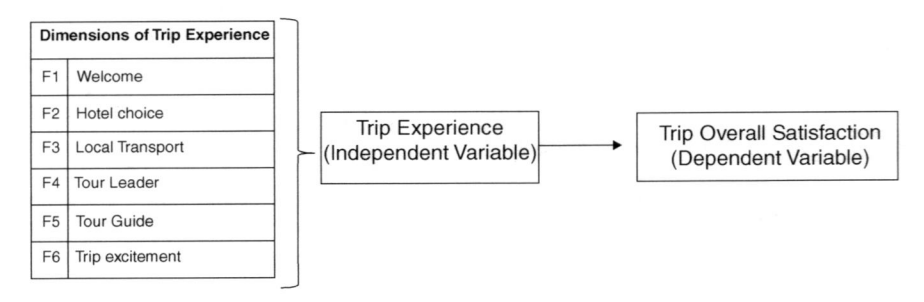

FIGURE 5.1 Proposed Conceptual Framework.

The study area chosen for the capital city of India, Delhi, has three World Heritage Sites, all of which fall under the cultural heritage category. According to Indian mythology, Delhi's history returned to Mahabharata's time. The documented history dates back to the 12th century, when the Delhi Sultanate was prominent. Thus, it was the capital city for many rulers in the past (Delhi Tourism 2022). In 2019, Delhi International Airport received the maximum number of tourists compared to all other international airports in India (Government of India 2020). In March 2020, Delhi Airport topped tourist arrivals, even under an e-tourist visa. Delhi is in fourth place for foreign tourist arrivals and thirteenth for domestic tourist arrivals. There was a growth of 8.86% in foreign tourists and 25.26% in domestic tourists from 2018 to 2019 (Government of India 2020).

Research methodology

Data were gathered using a questionnaire administered to the participants. A quantitative methodology was used to accomplish this objective. The acquired data were carefully analysed using statistical tests to investigate the links between the two variables. Two lists of attributes of trip experience and overall tourist satisfaction were compiled from literature. The trip experience attributes comprised 16 items, and the overall tourist satisfaction list included 5 items.

A questionnaire capturing the trip experience constructs, overall satisfaction constructs and demographic profiles of respondents was circulated among the tourists. Data analysis was conducted in two steps. A similar process was followed by Chi and Qu (2009). The initial step was to validate the variables for exploratory factor analysis (EFA) to identify the underlying dimensions of trip experience (Huete-Alcocer et al., 2019). Multiple regression analysis was performed using SPSS software to predict travellers' overall satisfaction based on six factors derived from the literature and the EFA (Chi & Qu, 2009).

Attributes of trip experience and overall satisfaction

A detailed literature review and unstructured discussions with a few tour operators generated a complete list of the trip experience and tourist satisfaction dimensions. While conducting the literature review, the characteristics of trip experience and tourist satisfaction were recorded and compiled into questionnaires reviewed by three senior tourism faculty members. Two lists of attributes of trip experience and overall tourist satisfaction were compiled from relevant tourism literature. The 16 items comprised the "trip experience" attributes; the overall tourist satisfaction list included five things.

After consulting the literature and talking to experts, the six trip experience attributes selected within each dimension varied greatly to reflect all the appropriate qualities. The first domain was welcoming, which included receiving services at the airport or station and transferring services to the hotel. Hotel choice was the second domain, which covered four features: room comfort, cleanliness and hygiene, food variety and exposure to new food. The third domain of local transport covers the vehicle's overall state and seat comfort.

The tour leader was the fourth attribute that included professional knowledge and interaction with tourists. Knowledge about the sites and communication skills were evaluated under the fifth head of the tour guide. The last attribute, trip excitement, was assessed regarding engagement, live shows, entertainment and shopping. The overall tourist satisfaction determinants were evaluated based on satisfaction with the trip, recommendations to others, social media posts, money spent and repeat visits. The questionnaire directed respondents to evaluate the tourist trip experience and satisfaction variables on a Likert scale of 1 to 5.

A pilot test of the research instrument was conducted to confirm its reliability and internal consistency using data collected from 30 tourists. An internal reliability test revealed Cronbach's alpha values of (0.823) for 16 'trip experience' items, and a score of (0.837) was observed for 5 items used for overall satisfaction. A score of 0.7 value, according to Nunnally and Bernstein (2010), is considered an acceptable value for instrument reliability.

A structured questionnaire, often used in tourist research, was used to gather the data administered by the participants (Hassan & Shahnewaz, 2014; Chi, 2012). The respondents were tourists who visited Delhi's three heritage sites. Over 450 questionnaires were distributed to 250 tourists at the three World Heritage Sites of Delhi: Red Fort, Humayun's Tomb and Qutb Minar and its monuments. The remaining 200 questionnaires were distributed across seven hotels.

The final questionnaire was distributed after conducting a pilot test with 30 hotel respondents (Hassan & Shahnewaz, 2014). The pilot test's result helped modify the final questionnaire by eliminating and rephrasing the questions the respondents needed to answer due to a lack of knowledge during their travel. Five questions were excluded from the final questionnaire. Data were collected using convenience sampling. Three hundred and nine were usable questionnaires (180 from hotels and 129 from sites) out of 450 questionnaires distributed.

Results

Tourists' demographic profiles are shown in Table 5.1. The response rate was 68.67%. As seen in Table 5.1, a major chunk of respondents (49%) fell in the 36–55 years category, while (42%) belonged to the 16–35 years. A higher percentage of the respondents were male (67%).

TABLE 5.1 Demographic profile of respondents

Background	Categories	Frequency	Percentage
Type of Tour Package			
	Group Tour Package	241	78
	Customised Tour Package	68	22
Age			
	16–35 years	131	42
	36–55 years	153	49
	Above 55 years	25	9
Gender			
	Female	102	33
	Male	207	67
Nationality			
	Indian	215	70
	Foreigner	94	30
Occupation			
	Student	108	35
	Employed/Self-Employed/Working at home	179	58
	Retired	22	7

Source: Authors' elaboration.

Exploratory factor analysis

Factor analysis summarises the observed data for a straightforward interpretation of relationships (Yong and Pearce, 2013). EFA reduces variables by exploring the dataset into small factors (Child, 2006). To confirm the predetermined dimensions of a trip experience and validate the variables for further analysis, the principal axis method of factor analysis was used as recommended by Chi and Qu (2009) and Huete-Alcocer, López-Ruiz and Grigorescu (2019). For this purpose, the dataset was closely examined, and outliers were removed before analysis.

Bartlett's test of sphericity was conducted to establish the correlations among the variables. The analysis revealed a significant Bartlett's score of .000 and a KMO-MSA value of the KMO-MSA (0.9), which were considered suitable for factor analysis. Eigenvalues, variance percentage, communality and factor-loading significance determined the extraction criteria of the factors. An eigenvalue of 1 or more was considered significant for element extraction. A total variance of 60% was considered substantial to ensure the significance of the derived factors, as recommended by Chi and Qu (2009). Six factors were generated with eigenvalues above 1.0, explaining an aggregate variance of 62.3% (see Table 5.2).

The value of commonalities was observed to be in a significant range (0.50 to 0.92), thereby confirming the variance in each of the six common factors

TABLE 5.2 Dimensions of the trip experience: Exploratory factor analysis

		Eigenvalue	Variance Explained	Cronbach's Alpha	Factor Loadings	Communality
F1	Welcoming	3.76	8.78	0.88		
	Receiving at the airport or station				0.71	0.69
	Transfer services to the hotel				0.69	0.67
F2	Hotel choice	8.45	18.96	0.78		
	Hygiene and cleanliness				0.82	0.78
	Exposure to new food.				0.88	0.77
	Variety of food				0.78	0.69
	Room comfort				0.8	0.71
F3	Local Transport	1.76	7.78	0.88		
	State of the vehicle				0.8	0.71
	Comfort of the seat				0.88	0.77
F4	Tour Leader	4.76	9.78	0.71	0.78	0.69
	Tour leader's knowledge.				0.8	0.71
	Tour leader's interaction skills.				0.78	0.69
F5	Tour Guide	4.76	8.78	0.71		
	Tour guide's knowledge				0.7	0.69
	Tour guides linguistic abilities.				0.8	0.71
F6	Trip excitement	6.76	9.62	0.73		
	Trip interest				0.71	0.69
	Trip engagement				0.8	0.71
	Evening entertainment				0.76	0.74
	Shopping				0.67	0.65

Source: Authors' elaboration.

computed together. In addition, the value of the 16-factor loadings was observed to be in the significant range between 0.45 and 0.84, that is greater than 0.35, considered an acceptable threshold for statistical significance. Thus, 16 variables loaded significantly on the 6 dimensions proposed in the study framework (welcoming, hotel choice, local transport, tour leader, tour guide, trip excitement). The reliability test score, Cronbach's alpha value for the six factors of the model, extended from 0.84 to 0.90 > 0.70, which is considered acceptable (Nunnally, 1978).

Multiple regression analysis

The subsequent step was to ascertain whether the identified attributes of the trip experience influenced overall tourist satisfaction. It was hypothesised that there would be no significant relationship between the trip experience and overall tourist satisfaction. Multiple regression analysis (MRA) was conducted to verify this hypothesis and explore the relative significance of each attribute of the trip experience contributing to tourists' overall satisfaction.

The multiple regression analysis predicted tourists' overall satisfaction comprising the mean score of (trip satisfaction, trip recommendation, social media post, money spent on the trip and repeat visit) as the dependent variable and the six attributes of trip experience as independent variables: welcoming (two items), hotel choice (four items), local transport (two items), tour leader (two items), tour guide (two items) and trip excitement (four items). Table 5.2 lists the 6 attributes and 16 measurement scales for these attributes. The values of the regression coefficients determined the corresponding importance of each factor (Table 5.3): trip excitement was found to be the heaviest factor ($\beta = 0.22$), followed by hotel choice ($\beta = 0.17$), tour guide ($\beta = 0.15$) and tour leader ($\beta = 0.14$). Thus, the final regression model is as follows:

$$Y = 1.69 + .18X1\,(\text{trip excitement}) + .14\,X2\,(\text{hotel choice}) + .12X3\,(\text{tour guide}) \\ + .06X4\,(\text{tour leader}) + .04X5\,(\text{welcoming}) + e$$

The results revealed that the five predictors together, $F\,(5,308) = 27.84$, $p < .05$, may account for 62.3% of the variance in tour experience. Table 5.3 shows the independent contributions of these predictors. The results showed that trip excitement ($\beta = 0.22$, $t = 3.37$, $p = 0.00$), hotel choice ($\beta = 0.17$, $t = 2.36$, $p = 0.02$), tour guide ($\beta = 0.15$, $t = 2.04$, $p = 0.04$), tour leader ($\beta = 0.14$, $t = 0.83$, $p = 0.02$) and welcoming ($\beta = 0.12$, $t = 0.53$, $p = 0.04$) positively predicted overall tourist satisfaction. The results suggest that trip excitement, hotel choice, tour guides, tour leaders and welcomes contribute to overall tourist satisfaction. In contrast, local transport made a minor contribution to the predictive power of the regression model.

Discussion of results

This study examines the trip experience as a multidimensional construct and examines the relationship between the various components of the trip

TABLE 5.3 Regression results of dimensions of trip experience on overall tourist satisfaction

Model		B	S.E.	β	t	P
1	(Intercept)	1.69	0.17		9.90	0.00
	Trip excitement	0.18	0.05	0.22	3.37	0.00
	Hotel choice	0.14	0.06	0.17	2.36	0.02
	Tour Guide	0.12	0.06	0.15	2.04	0.04
	Tour Leader	0.06	0.07	0.14	0.83	0.02
	Welcoming	0.04	0.04	0.12	0.53	0.04
	Local transport	0.05	0.05	0.08	0.91	0.61

Note: Dependent variable: overall tourist satisfaction.
Source: Authors' elaboration.

experience and overall tourist satisfaction. A two-step analysis was performed to accomplish the aim of the study. In the first step, exploratory factor analysis (EFA) was conducted on the 16 variables loaded on the 6 dimensions (welcoming, hotel choice, local transport, tour leader, tour guide and trip excitement) of trip experience. The findings showed that the 16 variables loaded significantly on the 6 dimensions proposed in the study framework. In the second step, multiple regression was used to predict tourists' overall satisfaction (trip satisfaction, trip recommendation, social media posts, money spent on the trip and repeat visits) on the six attributes of the trip experience. The results indicated that five attributes (trip excitement, hotel choice, tour guide, tour leader and welcoming) significantly affected overall tourist satisfaction. As a result of rejecting the null hypothesis, the study acknowledges a strong relationship between trip experience and overall tourist satisfaction.

This study supports the conclusions of earlier research by Chi and Qu (2009), who confirmed the positive impact of lodging, attractions, environment and dining as significant factors in tourist satisfaction. A study by Prebensen, Woo, Chen and Uysal (2012) support that the trip experience's dimensions help create the tourist's satisfaction with the destination visited. This study also confirms that trip experience positively affects tourist satisfaction (Kim & Brown, 2012; Altunel & Erkurt, 2015). These results agree with those (De Vos (2017) and Alrawadieh et al. (2019), who pointed out that trip experience influences tourist satisfaction.

Other related studies (Chen & Rahman, 2018; Bayih & Singh, 2020; Osman et al., 2020) have revealed that a memorable trip experience positively influences revisit intention and recommends a trip to others. In addition, quality experiences lead to recommending a trip to others (Altunel & Erkurt, 2015), and unforgettable experiences strongly impact revisit intentions (Barnes et al., 2016). Cetin (2020) correctly acknowledged that a positive tourist experience adds to a destination's popularity. The five dimensions highlighted by our study coincide with prior research (Alrawadieh et al., 2019; Su et al., 2020) that states these factors to be of prime concern and need improvement.

Conclusion

The role of tour operators is crucial in managing the tourist experience, from the reception of tourists to their departure. Our study examined the connection between the attributes of trip experience and overall tourist satisfaction. The study concluded that five attributes of trip experience (welcome, hotel, tour leader, tour guide and trip excitement) influenced tourist satisfaction in visiting Delhi. The findings revealed that tour operators can ensure tourist satisfaction by enhancing the five attributes of trip experience.

Previous arguments support the current findings of our study on the influence of trip experience on overall tourist satisfaction. Tour operators should work on the dimensions of the tourist experience, as it is pivotal for tourist satisfaction (Huang et al., 2010; Dahles, 2013; Cetin, 2020). Of these five dimensions, tour guides and leaders are important for tourist satisfaction (Cetin & Yarcan, 2017; Teng & Tsai, 2020; Weiler & Black, 2021). The remaining three dimensions of trip excitement, hotel choice and welcoming have been recognised by several other studies (Kim & Brown, 2012; Räikkönen & Honkanen, 2013; Liao & Chuang, 2020; Sthapit et al., 2019) as significant for tourist satisfaction. According to Alexander et al. (2020), a joyous trip experience can lead tourists to become ambassadors. Our study's theoretical and managerial implications focus on improving tourist experience to enhance tourist satisfaction.

Theoretical implications

Our research has successfully demonstrated that of the six trip experience attributes, five – trip excitement, hotel choice, tour guide, tour leader and welcoming – significantly impact the trip experience. These results are a major addition to the tourism research literature. Previous studies have looked into the attributes of trip experience and satisfaction separately, but this study explored the relationship between the two. This finding contributes to the literature by stating the relationship between the five dimensions of trip experience.

A literature review indicates that monitoring tourist satisfaction is essential for tourism (Mutanga et al., 2017). Satisfaction creates memorable experiences and a favourable outlook towards a holiday destination (Pestana et al., 2020). Tourist satisfaction indicates how well a product performs with respect to tourist expectations. Our study demonstrated that these five dimensions impact the overall satisfaction of the trip. Higher satisfaction leads to repeat tourists and destination recommendations (Valverde-Roda et al., 2022).

Managerial implications

This study will help tour operators enhance their trip experience by focusing on the five highlighted attributes of trip experience. The study proved that emphasising these five attributes would achieve overall tourist satisfaction. The first attribute that tour operators should consider is the hassle-free welcoming of tourists at airports and stations. Second, tour operators should ensure that the hotel room is comfortable and clean and that new gastronomical-related services are being met. Third, tour operators must ensure that the tour leader is knowledgeable and has good interactive skills. Fourth, the tour guide has good linguistic skills and is well informed about the heritage site, stimulating tourists' fascination with the site's historical and cultural significance. Fifth, tour operators must ensure that the trip is an enjoyable and refreshing experience for tourists, along with ample shopping time and good evening entertainment.

The financial success of tour operators is directly linked to their tourist satisfaction. The determinants of overall tourist satisfaction included satisfaction with the trip, recommendations to others, social media posts, repeat visits and money spent. Earlier studies have concluded that tourists whose expectations have been met by the services offered during their trip are satisfied and happy (Chen-Yi, 2016; Ghose & Johann, 2019). Tourists with memorable experiences will suggest the trip to others, share 'EWOM' and 'word of mouth', and repeat their visits (Hollebeek & Rather, 2019; Sthapit et al., 2019). The money spent on a trip affects tourist satisfaction (De Vos 2017). To have an important effect on the determinants of overall satisfaction, the five attributes of the trip experience should be managed well.

Limitations and future research

Despite the study's important contributions, some aspects can be improved. First, it focused on tourists choosing tour packages or customised tour operators. Independent tourists should be incorporated into future research. Second, this study was limited to Delhi, India. In contrast, studies of the same nature may be conducted for other tourist sites in India and abroad. Third, the questionnaire was distributed to the sample when they visited a heritage site or hotel. The data could be collected at the end of the trip to provide an overview of their entire experience for further study. Finally, other studies can assess which trip experience domain significantly influences tourist satisfaction, which can add to the next level of knowledge.

References

Alexander, K. A., Liggett, D., Leane, E., Nielsen, H. E. F., Bailey, J. L., Brasier, M. J., & Haward, M. (2020). What and who is an Antarctic ambassador? *Polar Record*, *55*(6), 497–506. https://doi.org/10.1017/S0032247420000194

Alrawadieh, Z., Alrawadieh, Z., Kozak, M. (2019a). Exploring the impact of tourist harassment on destination image, expenditure, and loyalty. *Tourism management*, *73*, 13–20. https://doi.org/10.1016/j.tourman.2019.01.015

Alrawadieh, Z., Prayag, G., Alrawadieh, Z., & Alsalameen, M. (2019b). Self-identification with a heritage tourism site, visitors' engagement, and destination loyalty: The mediating effects of overall satisfaction. *The Service Industries Journal*, *39*(7–8), 541–558. https://doi.org/10.1080/02642069.2018.1564284

Altunel, M. C., & Erkurt, B. (2015). Cultural tourism in Istanbul: The mediation effect of tourist experience and satisfaction on the relationship between involvement and recommendation intention. *Journal of Destination Marketing & Management*, *4*(4), 213–221. https://doi.org/10.1016/j.jdmm.2015.06.003

Arnould, F.L., & Price, L.L. (1993). River magic: Extraordinary experiences and extended service encounters. *Journal of Consumer Research*, *22*, 24–45. https://doi.org/10.1086/209331

Barnes, S.J., Mattsson, J., Sørensen, F. (2016). Remembered experiences and visit intentions: A longitudinal study of Safari park visitors. *Tourism Management*, *57*, 286–294. https://doi.org/10.1016/j.tourman.2016.06.014

Barnes, S. J., Mattsson, J., Sørensen, F., and Friis Jensen, J. (2020). The mediating effect of experiential value on tourist outcomes from encounter-based experiences. *Journal of Travel Research*, *59*(2), 367–380. https://doi.org/10.1177/0047287519837386

Basuroy, T. (2022). Internet usage in India: statistics and facts. Retrieved from https://www.statista.com/topics/2157/internet-usage-in-india/#topicOverview

Bayih, B. E., & Singh, A. (2020). Modeling domestic tourism: motivations, satisfaction, and tourist behavioral intentions. *Heliyon*, *6*(9), e04839. https://doi.org/10.1016/j.heliyon.2020.e04839

Bergadaa, M. and Lorey, T. (2015). Preservation of living cultural heritage: The case of basque choirs and their audiences. *International Journal of Art Management*, *17*(3), 15.

Cajiao, D., Leung, Y. F., Larson, L. R., Tejedo, P., & Benayas, J. (2022). Tourists' motivation, learning, Marinao, and trip satisfaction facilitate the pro-environmental outcomes of the Antarctic tourist experience. *Journal of Outdoor Recreation and Tourism*, *37*, 100454. https://doi.org/10.1016/j.jort.2021.100454

Carbone, L. & Haeckel, S. (1994). Engineering customer experience. *Marketing Management*, *3*, 8–19.

Cetin, G. (2020). Experience vs quality: predicting satisfaction and loyalty in services. *The Service Industries Journal*, *40*(15–16), 1167–1182. https://doi.org/10.1080/02642069.2020.1807005

Cetin, G., & Bilgihan, A. (2015). Components of cultural tourists' experiences in destinations. *Current Issues in Tourism*, *19(2)*, 137–154. https://doi.org/10.1080/13683500.2014.994595

Cetin, G., & Yarcan, S. (2017). The professional relationship between tour guides and tour operators. *Scandinavian Journal of Hospitality and Tourism*, *17*(4), 345–357. https://doi.org/10.1080/15022250.2017.1330844

Chang, T. Y., Hung, S. F., & Tang, S. (2022). Seek common ground local culture while reserving differences: Exploring types of souvenir attributes by Ethnic Chinese people. *Tourist Studies*, *22*(1), 21–41. https://doi.org/10.1177/14687976211035961

Chen, H., & Rahman, I. (2018). Cultural tourism: An analysis of engagement, cultural contact, memorable experience, and destination loyalty. *Tourism Management Perspectives*, *26*(1), 153–163. https://doi.org/10.1016/J.TMP.2017.10.006

Chen-Yi, W. (2016). Tourist behavioral intentions and festival quality: The case of Kaohsiung Lantern festival. *International Journal of Research in Tourism and Hospitality*, *2*(3), 23–28. ISSN 2455–0043.

Chi, C., & Qu, H. (2009). Examining the relationship between tourists' attribute satisfaction and overall satisfaction. *Journal of Hospitality Marketing and Management*, *18*(1), 4–25. https://doi.org/10.1080/19368620801988891

Chi, C.G. (2012). An examination of destination loyalty: differences between first-time and repeat visitors. *Journal of Hospitality and Tourism Research*, *36 (1)*, 3–24. https://doi.org/10.1177/1096348010382235

Child, D. (2006). *The essentials of factor analysis*. (3rd ed.). Continuum International Publishing Group.

Dahles, H. (2013). *Tourism, heritage, and national culture in Java: Dilemmas of a local community*. Routledge.

De Vos, J. (2017). Analyzing the effect of trip satisfaction on satisfaction with leisure activity at the trip's destination about life satisfaction. *Transportation*, *46*, 623–645. https://doi.org/10.1007/s11116-017-9812-0

Delhi Tourism. (2022). *About delhi*. Retrieved from http://delhitourism.gov.in/delhitourism/aboutus/index.jsp

Devaraja, T. S., & Deepak, K. (2018). Relationship between tour operators and tourists towards tourism development in India: A study on the Mysore District, Karnataka. *Journal of Hotel and Business Management*, *7*(2), 2–8. https://doi.org/10.4172/2169-0286.1000187

Dieck, C., & Jung, T. (2018). A theoretical model of mobile Augmented Reality acceptance in urban heritage tourism. *Current Issues in Tourism*, *21*(18), 154–174. https://doi.org/10.1080/13683500.2015.1070801

Domínguez-Quintero, A. M., González-Rodríguez, M. R., & Paddison, B. (2020). The mediating role of experience quality on authenticity and satisfaction in cultural heritage tourism. *Current Issues in Tourism*, *23*(2), 248–260. https://doi.org/10.1080/13683500.2018.1502261

Elghani, M. (2012). *Heritage and hospitality links in hotels in Siwa, Egypt: Towards providing authentic experiences*. (Doctoral dissertation, University of Waterloo, Ontario, Canada). Retrieved from http://hdl.handle.net/10012/7074

Fairhurst, A., Costello, C., & Holmes, A. (2007). An examination of the shopping behavior of visitors to Tennessee according to tourist typologies. *Journal of Vacation Marketing*, *13*(4), 311–320. https://doi.org/10.1177/1356766707081005

Fang, S., Zhang, C., & Li, Y. (2020). Physical attractiveness of service employees and customer engagement in the tourism industry. *Annals of Tourism Research*, *80*, 102756. https://doi.org/10.1016/j.annals.2019.102756

Fuentes-Moraleda, L., Díaz-Pérez, P., Orea-Giner, A., Muñoz-Mazón, A., & Villacé-Molinero, T. (2020). Interaction between hotel service robots and humans: A hotel-specific Service Robot Acceptance Model (sRAM). *Tourism Management Perspectives*, *36*, 100751.

Garau, C. (2017). Emerging technologies and cultural tourism: Opportunities for a cultural urban tourism research agenda. *Tourism in the City*, 67–80. https://doi.org/10.1007/978-3-319-26877-4_4

Garcia, M. (2023). Consumer buying behavior on travel products through online travel agencies. *SSRN* 4442816. http://doi.org/10.2139/ssrn.4442816

Ghose, S., & Johann, M. (2019). Measuring tourist satisfaction with destination attributes. *Journal of Management and Financial Sciences*, *34*, 9–22. https://doi.org/10.33119/JMFS.2018.34.1

Government of India, Ministry of Tourism. (2020). *India Tourism Statistics, 2020.* New Delhi: Marketing Research Division. Retrieved from https://tourism.gov.in/market-research-and-statistics

Gračan, D., Zadel, Z., & Pavlović, D. (2021). Management of visitor satisfaction using mobile digital tools and services to create a smart destination. *Ekonomski pregled*, *72*(2), 185–198. https://doi.org/10.32910/ep.72.2.2

Hassan, M.M., & Shahnewaz, Md. (2014). Measuring tourist service satisfaction at the destination: A case study of Cox's Bazar sea beach in Bangladesh. *American Journal of Tourism Management*, *3*(1), 32–43. https://doi.org/10.5923/j.tourism.20140301.04

Hollebeek, L., & Rather, R. (2019). Service innovativeness and tourism customer outcomes. *International Journal of Contemporary Hospitality Management*, *31*(11), 4227–4246. https://doi.org/10.1108/IJCHM-03-2018-0256

Huang, S., Hsu, C. H. C., & Chan, A. (2010). Tour guide performance and tourist satisfaction: A study of package tours in Shanghai. *Journal of Hospitality and Tourism Research 34*(1), 3–33. https://doi.org/10.1177/1096348009349815

Huete-Alcocer, N., López-Ruiz, V.R., & Grigorescu, A.(2019). Measurement of satisfaction with sustainable tourism: A cultural heritage site in Spain. *Sustainability*, *11*(23), 67–74. https://doi.org/10.3390/su11236774

Io, M. U. (2019). Understanding the core attractiveness of performing arts heritage to international tourists, *Tourism Geographies*, *21*(4), 687–705. https://doi.org/10.1080/14616688.2019.1571096

Kaur, A., Chauhan, A., Medury, Y. (2016). Destination image of Indian tourism destinations: An evaluation using correspondence analysis. *Asia Pacific Journal of Marketing and Logistics*, *22*(3), 499–524. https://doi.org/10.1108/APJML-05-2015-007

Kim, A., & Brown, G. (2012). Understanding the relationships between perceived travel experiences, satisfaction, and destination loyalty. *Anatolia*, *23*(3), 328–347. https://doi.org/10.1080/13032917.2012.696272

Kim, J.H., Ritchie, J.R.B., & McCormick, B. (2012). Development of a scale to measure memorable tourism experiences. *Journal of Travel Research*, *51*(1), 12–25. https://doi.org/10.1177/0047287510385467

Kong, W.H. & Chang, T. (2016). Souvenir shopping, tourist motivation, and travel experience. *Journal of Quality Assurance in Hospitality & Tourism*, *17*(2), 163–177. https://doi.org/10.1080/1528008X.2015.1115242

Lee, J., Kyle, G., & Scott, D. (2012). The mediating effect of place attachment on the relationship between festival satisfaction and loyalty to host destinations. *Journal of Travel Research*, *51*(6), 754–767. https://doi.org/10.1177/0047287512437859

Lee, W., & Chhabra, D. (2015). Heritage hotels and historic lodging: Perspectives on experiential marketing and sustainable culture. *Journal of Heritage Tourism*, *10*(2), 103–110. https://doi.org/10.1080/1743873X.2015.1051211

Lee, Y.-G., Yim, B. H., Jones, C. W., & Kim, B. G. (2016). The extended marketing mix in the context of dance is performing art. *Social Behavior and Personality: An International Journal*, *44*(6), 1043–1056. https://doi.org/10.2224/sbp.2016.44.6.1043

Liao, C.-S. Chuang, H.-K. (2020). Tourist preferences for package tour attributes in tourism destination design and development. *Journal of Vacation Marketing*, *26*(2), 230–246. https://doi.org/10.1177/1356766719880250

Loncaric, D., Dlacic, J., & Bagaric, L. (2019). Exploring the relationship between satisfaction with tourism services, revisit intention, and life satisfaction. *41st International Scientific Conference on Economic and Social Development* (pp. 122–132). Belgrade.

Marin-Pantelescu, A., Tăchiciu, L., Căpuşneanu, S., & Topor, D.I. (2019). Role of tour operators and travel agencies in promoting sustainable tourism. *Amfiteatru Economic*, *21*(52), 654–669. https://doi.org/10.24818/EA/2019/52/654

Mawufemor, K., Eshun, G., & Tichaawa, T. M. (2019). Factors influencing the choice of souvenirs by international tourists. *African Journal of Hospitality, Tourism and Leisure*, *8*(5), 1–10. https://www.ajhtl.com/uploads/7/1/6/3/7163688/article_54_vol_8_5__2019_uj.pdf

Mora, D., Solano-Sánchez, M. Á., López-Guzmán, T., & Moral-Cuadra, S. (2021). Gastronomic experience is a key element in the development of tourist destinations. *International Journal of Gastronomy and Food Science*, *25*, 100405. https://doi.org/10.1016/j.ijgfs.2021.100405

Morris, E.A., & Guerra, E. (2015). Are we there yet? Trip duration and mood during travel. *Transportation Research Part F: Traffic Psychology and Behaviour*, *33*, 38–47. https://doi.org/10.1016/j.trf.2015.06.003

Mrsic, L., Surla, G., & Balkovic, M. (2020). Technology-driven smart support systems for tourist destination management organizations. In A. Khanna, D. Gupta, S. Bhattacharyya, V. Snasel, J. Platos, & A. Hassanien (Eds.), *International Conference on Innovative Computing and Communications. Advances in Intelligent Systems and Computing* (pp. 65–76). Springer.

Murphy, P., Pritchard, M., & Smith, J. (2000). The destination product and its impact on traveler perceptions. *Tourism Management*, *21*, 43–52. https://doi.org/10.1016/S0261-5177(99)00080-1

Mutanga, C. N., Vengesayi, S., Chikuta, O., Muboko, N., & Gandiwa, E. (2017). Travel motivation and tourist satisfaction with wildlife tourism experiences in Gonarezhou and Matusadona National Parks, Zimbabwe. *Journal of Outdoor Recreation and Tourism*, *20*, 1–18. https://doi.org/10.1016/j.jort.2017.08.001

Nunnally, J. C. (1978). *Psychometric theory* (2nd ed.). McGraw-Hill.

Nunnally, J. C., & Bernstein, I. (2010). *Psychometric theory* (3rd ed.). Tata McGraw-Hill Ed.

Oliveira, T., Araujo, B., & Tam, C. (2020). Why do people share their travel experiences on social media? *Tourism Management*, *78*, 104041. https://doi.org/10.1016/j.tourman.2019.104041

Oliver, R. L. (1993). A conceptual model of service quality and satisfaction: Compatible goals and concepts. In T. A. Swartz, D. E. Bowen, & S. W. Brown (Eds.), *Advances in services marketing and management* (Vol. 2, pp. 65–85). JAI Press.

Osman, H., Brown, L., & Phung, T. M. T. (2020). Travel motivations and experiences of female Vietnamese solo travelers. *Tourist Studies*, *20*(2), 248–267. https://doi.org/10.1177/1468797619878307

Page, S. J., & Connell, J. (2020). *Tourism: A modern synthesis* (5th ed.). Routledge.

Pai, C. K., Liu, Y., Kang, S., & Dai, A. (2020). The role of perceived smart tourism technology in tourist satisfaction, happiness, and revisit intention. *Sustainability*, *12*(16), 6592. https://doi.org/10.3390/su12166592

Park, C., & Petrick, J. F. (2016). Developing an optimal Korean performing arts tourism product for Japanese tourists. *Tourism Travel and Research Association: Advancing Tourism Research Globally, 16*, 1–7.

Pestana, M. H., Parreira, A., Moutinho, L. (2020). Motivations, emotions, and satisfaction: Key to tourism destination choice. *Journal of Destination Marketing & Management, 16*, 100332. https://doi.org/10.1016/j.jdmm.2018.12.006

Pourabedin, Z., Mahony, T., Pryce, J. (2022). To develop a multisensory scale to capture the attributes of heritage boutique hotels. In T. Chaiechi, & J. Wood (Eds.), *Community empowerment, sustainable cities, and transformative economies* (pp. 355–375). Springer.

Prebensen, N. K., Woo, E., Chen, J. S., & Uysal, M. (2012). Experience quality in different phases of a tourist vacation: A case in northern Norway. *Tourism Analysis, 17*(5), 617–627. https://doi.org/10.3727/108354212X13485873913921

Räikkönen, J., & Honkanen, A. (2013). Does satisfaction with package tours lead to successful vacation experiences? *Journal of Destination Marketing and Management, 2*(2), 108–117. https://doi.org/10.1016/j.jdmm.2013.03.002

Ray, S. (2022). Technology and destination promotion in Asian tourism: Challenges, changes, and bearing. In A. Hassan (Ed.), *Handbook of technology application in tourism in Asia* (pp. 377–402). Springer.

Sangpikul, A. (2018). The effects of travel experience dimensions on tourist satisfaction and destination loyalty: The case of an island destination. *International Journal of Culture, Tourism and Hospitality Research, 12*(2). https://doi.org/10.1108/IJCTHR-06-2017-0067

Santa-Cruz, F. G., & López-Guzmán, T. (2017). Culture, tourism & world heritage sites. *Tourism Management Perspectives, 24*, 111–116. https://doi.org/10.1016/j.tmp.2017.08.004

Selim, H., Eid, R., Agag, G., & Shehawy, Y. M. (2022). Cross-national differences in travelers' continuance of knowledge sharing in online travel communities. *Journal of Retailing and Consumer Services, 65*, 102886. https://doi.org/10.1016/j.jretconser.2021.102886

Sthapit, E., Del Chiappa, G., Coudounaris, D. N. & Bjork, P. (2019). Tourism experiences, memorability, and behavioral intentions: A study of tourists in Sardinia, Italy. *Tourism Review, 75*(3), 533–558. https://doi.org/10.1108/TR-03-2019-0102

Su, D. N., Nguyen, N. A. N., Nguyen, Q. N. T., & Tran, T. P. (2020). The link between travel motivation and satisfaction with a heritage destination: The role of visitor engagement, visitor experience, and heritage destination image. *Tourism Management Perspectives, 34*, 100634. https://doi.org/10.1016/j.tmp.2020.100634

Su, L., Chen, H., & Huang, Y. (2022). Tourists' monetary and temporal sunk costs influence their destination trust and visit intentions. *Tourism Management Perspectives, 42*, 100968. https://doi.org/10.1016/j.tmp.2022.100968

Swarbrooke, J. (2001). Key challenges for visitor attraction managers in the U.K. *Journal of Leisure Property, 1*(4), 318–336.

Syakier, W. A., & Hanafiah, M. H. (2022). Tour guide performances, tourist satisfaction, and behavioral intentions: A study on Kuala Lumpur City Center tours. *Journal of Quality Assurance in Hospitality & Tourism, 23*(3), 597–614. https://doi.org/10.1080/1528008X.2021.1891599

Taheri, B., Hosany, S., & Altinay, L. (2019). Consumer engagement in the tourism industry: new trends and implications for research. *Service Industries Journal, 39*(7–8), 463–468. https://doi.org/10.1080/02642069.2019.1595374

Teng, H., & Tsai, C. (2020). Can tour leaders' likability enhance tourists' value co-creation behavior? The role of attachment. *Journal of Hospitality and Tourism Management*, *45*, 285–294. https://doi.org/10.1016/j.jhtm.2020.08.018

Ugwuanyi, C.C., Ehimen, S. & Uduji, J.I. (2021). Hotel guests' experience, satisfaction, and revisit intentions: An emerging market perspective. *African Journal of Hospitality, Tourism and Leisure*, *10*(2), 406–424. https://doi.org/10.46222/ajhtl.19770720-108

Valverde-Roda, J., Moral-Cuadra, S., Aguilar-Rivero, M., & Solano-Sánchez, M. Á. (2022). Perceived value, satisfaction, and loyalty at the world heritage sites Alhambra and Generalife (Granada, Spain). *International Journal of Tourism Cities*, *8*(4), 946–964. https://doi.org/10.1108/IJTC-08-2021-0174

Weiler, B., & Black, R. (2021). The changing face of the tour guide: one-way communicator to choreographer to co-creator of the tourist experience. In G. Thi Phi, & D. Dredge (Eds.) *Critical Issues in Tourism Co-Creation* (1st ed., pp. 91–105). Routledge.

Weiler, B., & Yu, X. (2007). Dimensions of cultural mediation in guiding Chinese tour groups: implications for interpretation. *Tourism Recreation Research*, *32*(3), 13–22. https://doi.org/10.1080/02508281.2007.11081535

Wilkins, H. (2011). Souvenirs: What and why do we buy? *Journal of Travel Research*, *50*(3), 239–247. https://doi.org/10.1177/0047287510362782

Wu, H.-C., & Li, T. (2017). A study of experiential quality, perceived value, heritage image, experiential satisfaction, and behavioral intentions for heritage tourists. *Journal of Hospitality & Tourism Research*, *41*(8), 904–944. https://doi.org/10.1177/1096348014525638

Yameen, M. (2013). Marketing strategies for the Indian hotel industry. *Global Journal of Arts & Management*, *3*(3), 147–150.

Yen, C., Tsaur, S., & Tsai, C. (2018). Tour leaders' job crafting: Scale development. *Tourism Management*, *69*, 52–61. https://doi.org/10.1016/j.tourman.2018.05.017

Yong, A. G., Pearce, S. (2013). A beginner's guide to factor analysis: Focus on exploratory factor analysis. *Tutorials in Quantitative Methods for Psychology*, *9*(2), 79–94. https://doi.org/10.20982/tqmp.09.2.p079

Yoon, Y., & Uysal, M. (2005). Examination of the effects of motivation and satisfaction on destination loyalty: A structural model. *Tourism Management*, *26*(1), 45–56. https://doi.org/10.1016/j.tourman.2003.08.016

Zhang, G., Chen, X., Law, R., & Zhang, M. (2020). Sustainability of heritage tourism: A structural perspective from cultural identity and consumption intention. *Sustainability*, *12*(21), 9199. https://doi.org/10.3390/su12219199

Zhang, J., Xiong, K., Liu, Z., & He, L. (2022). Research progress and knowledge system of world heritage tourism: A bibliometric analysis. *Heritage Science*, *10*(1), 1–18. https://doi.org/10.1186/s40494-022-00654-0

Zhou, G., Liu, Y., Hu, J., & Cao, X. (2023). The effect of tourist-to-tourist interaction on tourists' behavior: The mediating effects of positive emotions and memorable tourism experiences. *Journal of Hospitality and Tourism Management*, *55*, 161–168. https://doi.org/10.1016/j.jhtm.2023.03.005

6

VISITORS' CONSUMPTION EXPERIENCE VIS-À-VIS VISITORS' REVISIT TO STREET FOOD MARKETS IN A POST-COVID ERA

A case study of Windhoek (Namibia)

Elsie Vezemburuka Hindjou, Ethilde Tulimuwo Kuwa and Saurabh Kumar Dixit

Introduction

Food preparation is a significant aspect of identity for many destinations. As a result, culinary tourism offers a chance to re-energise and diversify tourism and support regional economic growth (UNWTO, 2017). Due to its blend of local cuisine and the destination's cultural significance, street food (SF) has grown in appeal as gastronomy tourism prospers (Jeaheng et al. 2023). FAO defines street food as

> food made from food and drinks prepared for consumption, prepared and sold on the street or in other public places such as markets and/or exhibitions, often sold on counters or through temporary vans and carts of street vendors.
>
> *(FAO, 2009, p. 7)*

Street food is becoming more popular all over the world and is particularly popular in several cities on continents, including Africa, India, Asia and Latin America (Imathiu, 2017). Despite its populace, there is a widespread belief that foods sold on the street are dangerous, exposed to various potential contaminants and mainly cooked and consumed in open spaces (Verma et al. 2023). Due to this perception of a significant public health risk, most consumers avoid street foods (Aziz & Dahan, 2013; Khairuzzaman et al. 2014). Reduced visits to street food markets can result from a fear of contagions, especially in the post-COVID era.

DOI: 10.4324/9781032637778-7

Street food/food tourism: Namibian current context

Food tourism increases diversity by allowing tourists to experience authentic and traditional local food at destinations (Birch & Memery, 2020), and Namibia is no exception. However, it is more complicated in Namibia, as it hosts 11 ethnic groups. No particular dish can, therefore, be singled out; rather, the most favoured dishes with influences from countries like South Africa.

There are approximately 11 different ethnic groups in Namibia, and each one has influenced the distinctiveness of genuine Namibian cuisine. One could define *authentic Namibian cuisine* as an infusion of European and African ingredients and recipes. On the European side, influences include those from Germany, Holland and England. On the Namibian side, inspirations include those from tribes like the Namas, Ovaherero, Damara and Ovambo (Ahrens & Ahrens, 2017). Namibia is an arid country, which means that traditional cuisines depend on seasonal ingredients that are endemic to particular rural areas. Namibia provides opportunities for visitors to sample some authentic Namibian cuisines. Due to the seasonality of such unique local ingredients, as well as mainly depending on the climate in the area, this food sampling offers a genuinely unique Namibian experience (Ahrens & Ahrens, 2017).

Considered a street food, Kapana is favoured all over the country and is enjoyed by both locals and tourists alike, making it a suitable case study for this chapter. Kapana is a popular street food in Namibia that is made from grilled beef or game meat. It is a quick and affordable meal, typically prepared by grilling thin slices of beef or game meat over an open flame or hot coals. The meat is seasoned with spices like salt, pepper and chilli powder, which add flavour and heat to the dish. Once cooked, the meat is typically served with a side of chilli sauce or salsa and eaten with the hands. One of the most famous street food markets where one can eat Kapana in the capital city of Namibia, Windhoek, is the Oshetu market (also known as single quarters or Kapana Market) in Katutura township.

Currently, Namibia is not strongly perceived as a food destination; however, there are aspects which can be developed to be a food-associated destination. Nevertheless, there is a need for cooperation amongst destination planners and other food tourism stakeholders to integrate local cuisines into Namibia's tourism marketing campaigns.

Food experience and visitors' satisfaction

Ozcelik and Akova (2021, p. 4176) define experience 'as a process which impresses consumers emotionally, spiritually and physically after certain consumption'. Food experience is one of the most critical factors affecting tourists' levels of contentment when travelling. According to Hendijani (2016), food experience is crucial to a tourist vacation because it may be a means to learn about culture and can be a source of joy. Everett and Aitchison (2008) also added that cuisine is critical in drawing tourists to destinations. Similarly, Jeaheng and Han (2020) agreed that food is essential to visiting Thailand as their cuisine reflects local people's cultural way of living by offering memorable and unique experiences to tourists. Cusack (2000), however, added that since food is acknowledged as a means of expressing culture and identity, it is crucial to cultural tourism. Most earlier studies show that visitors travel to experience and learn about other customs and cultures.

Although food experience is vital in promoting a destination, culture and traditions, visitor satisfaction is a crucial aspect of the overall tourist travel experience. Despite the importance of street food to the destination and its contribution to tourism, Ozcelik and Akova (2021) state that there needs to be more studies directed to the concept of the street food experience. Ozcelik and Akova (2021), in their studies about the impacts of street food experience and behavioural intentions, mentioned that understanding visitors' intentions to revisit specific locations has become crucial for a destination.

For destination management to maintain and preserve traditional street food to attract and please more tourists, it is vital to realise that eating on the street impacts behavioural intentions. For tourists to experience street food, health and hygiene are vital factors affecting behavioural intention. When tourists travel, they try out street food and drinks to experience a host community's cuisine. Thus, according to Ozcelik and Akova (2021), a host community's food and beverage businesses can enhance the tourists' experience by reinforcing regional identity and sense of locality. Particularly, the dining experience of tourists, which affects their happiness or discontent, is a crucial component in their decisions to revisit the place. Street food is a popular tourist attraction, and eating it significantly impacts one's behaviour.

Street food safety

While street foods are attractive sources of meals for both locals and visitors alike, they are also often perceived to be a significant public health risk due to unsafe food handling (Gemeda et al. 2023), more so in the third world and developing countries (Khairuzzaman et al. 2014; Samapundo et al. 2016). How some vendors store and prepare the food can promote cross-contamination, which poses a health risk (Ekka, 2017). As such, the microbiological quality of street-vended food is a paramount concern that significantly contributes to foodborne diseases (Sousa et al. 2022; Rosales et al. 2023).

According to the FAO and WHO (2022) report, the health risks associated with street food can be attributed to the lack of knowledge among street food vendors. Several studies share the above sentiments on the lack of adequate food safety attitudes, knowledge and practices among vendors due to the deficit of food hygiene training (Hiamey & Hiamey, 2018; Oladipo-Adekeye &

Tabit, 2021; Samapundo et al. 2015; Monney et al. 2013). Consequently, ensuring food safety and good hygiene practices is among the most formidable challenges the food vendors face daily (Khuluse & Deen, 2020).

In Windhoek, street vendors are typically clustered in overcrowded areas, including residential areas, near workplaces, open markets and busy street pavements. Given the areas in which they operate, they are likely to refrain from executing food safety procedures during food handling and preparation (Noor, 2016), which could lead to food contamination.

Methodology

An analysis of secondary data from relevant published articles and journals on food tourism, street food experience and safety was made and informed the literature review. This study also adopted netnography in the form of non-participant observation and is based on online reviews published on TripAdvisor (www.TripAdvisor.com) regarding street food consumption experiences and revisits to the street food market in Windhoek. This approach was taken to avoid influencing the reviews in any way, for the reviews to be as natural as possible (Kozinets, 2010). Furthermore, the passive netnographic approach applied in this study supports a high personal and social distance between researchers and bloggers (Arsal et al., 2010). Kozinets (2010, p. 89) offers six criteria when selecting sites for netnographic research: they should be 'relevant, active, interactive, substantial, heterogeneous, and data-rich'.

The keyword 'kapana' was used to extract user-generated text reviews written by visitors in English between 2011 and 2023 (prior to and post-COVID). The reviews extracted were posted by local and international visitors alike. The reviews represent international markets: South Africa, the Netherlands, Ireland, the United Arab Emirates, Australia, the United States, the United Kingdom, Austria and Brazil. The keyword 'kapana' was selected to retrieve reviews indicating elements influencing visitors' Kapana consumption experience. The data retrieval from TripAdvisor was conducted from 7 to 9 June 2023. The extracted reviews were further screened to remove those unrelated to the study's objective. These extracted data from TripAdvisor were read many times to understand the content of the reviews that visitors to the Kapana market posted. The thorough reading was followed by coding the data into topics to achieve the set objective. The data were further analysed and classified into themes; these themes were further analysed supported by examples of reviews posted by visitors. Those themed reviews deemed dominant were analysed; these dominant themed reviews are presented in the findings to provide more insights and a sense of what elements may influence visitor experience at the street food market.

A total of 41 online reviews were considered for the analysis. A similar approach has been adopted in other studies (Gunasekar et al. 2021) to capture customer satisfaction. In the next step of data processing, we used Taguette (a free and open-source qualitative research tool) to identify the factors and determine the extent to which each factor influences visitors' experience at the street food market in Windhoek were referenced. The coding process identified nine factors from the reviews referenced by visitors [Atmosphere, Authenticity, Cleanliness, Engaging with others, Fresh Meat, Fun, Safety, and Taste] as key factors influencing their experience. The extent was determined by the frequency of the factor(s) identified. Data were read and reviewed independently by each researcher to ensure that the information was related and consistent to the emerging themes.

Additionally, a survey of street food market visitors was conducted for empirical evidence (the survey responses consisted of locals visiting the market only. The factors identified from the online reviews were used as basis for the survey questions development. The survey was conducted between 1 July and 1 August 2023. Of the 60 questionnaires administered only 40 were considered for analysis, the 20 were incomplete and therefore excluded from the analysis. These formed the basis for understanding the effects of street food consumption experiences on visitors' revisits post-COVID. The survey focused on the relationship linking street food consumption experiences to visitors' market revisits. The data were manually entered into an Excel workbook; then the PivotChart was used to calculate the percentages to determine the satisfaction level and perceptions on Kapana.

Findings and discussion

The following narratives show guests' expressions concerning the elements influencing their experiences at the Kapana meat market in Windhoek.

If you love meat and i mean MEAT, you should visit the 'single quarters' in Katutura, almost everyone in Windhoek knows where it is! Chopped up steaks like never before (Kapana). Do note that it isn't that safe to go there alone if you're a foreigner, best to go with someone who knows the place well enough. Enjoy our townships meat that you'll never taste anywhere else! An experience you won't forget!

The best part is 'Kapana', where you get it at the open market. Kapana is fresh meat that is grilled and cut into small cubes for easy eating. It is eaten with kapana spice, chilli and salsa (tomato and onion). It is a must do, no matter your back-ground, unless of-course you are not a meat lover. The open market is set up for you to have an all rounder experience. Where you get you kapana, and you can get traditional bread or porridge(pap), salsa and soft drinks to finish off your meal. Your meal is served on clean tables and the environment set up for is clean and safe.

Visiting Windhoek without checking out Katutura? You won't have a story to tell. Come and experience Kapana in single quarter open market. Here you'll find all in one. You can buy your traditional food to take with home or just buy to taste. Come with a local person and enjoy. Or maybe just call me do i can help you eat Kapana. Haha

This place unlike townships in other countries is safe to explore and engage with the local population. There are several places to catch a drink and enjoy time mingling and often there is a car wash or braai spot very close to you. Many places serve the local food. Try out the local Kapana. Worth a try while there.

I've visited Windhoek for work on several occasions, however, this tour showed me areas of the city (and the city's history) that I hadn't seen before! A highlight was having Kapana in Katatura!

My favorite part was Kapana, a local delicacy which consist of 100% pure beef grilled on an open flame, served with Kapana spice, which I would highly recommend. As well as some of the highly nutritious traditional meals such as mopani worms. Asser took in the depth of Katutura to ensure we also had a chance to interact with the locals.

Taste was referenced as the highest factor influencing visitor experience in both the online reviews and the empirical survey, as shown in Figure 6.1. According to Cho et al. (2014), repeat visitors often think of food quality, which includes taste and freshness; this is due to the fact they perceive food as an essential item for their tourist experience that will increase their intention to revisit the said destination (Jayarman et al. 2010). Food quality is the quality characteristics of food that are acceptable to consumers (Kapiris, 2012); food quality attributes include freshness, menu variety, presentation, temperature, being healthy and tastefulness (Rozekhi et al. 2016). Atikahambar et al. (2018) further note that food quality influences destination choice in relation to experience.

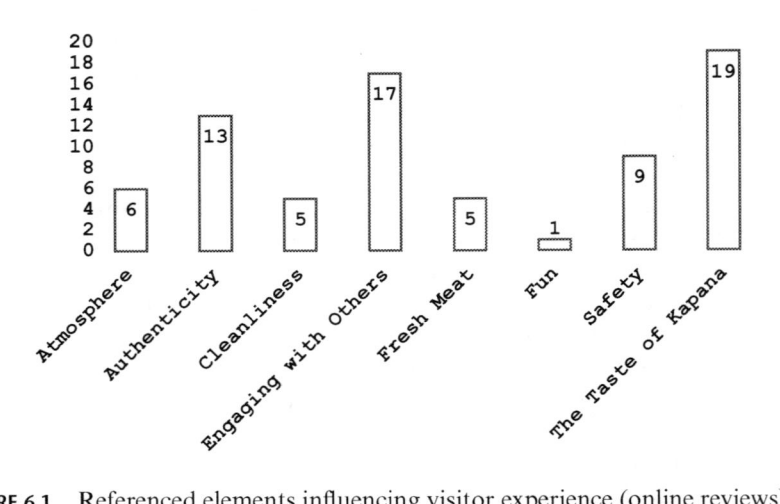

FIGURE 6.1   Referenced elements influencing visitor experience (online reviews).

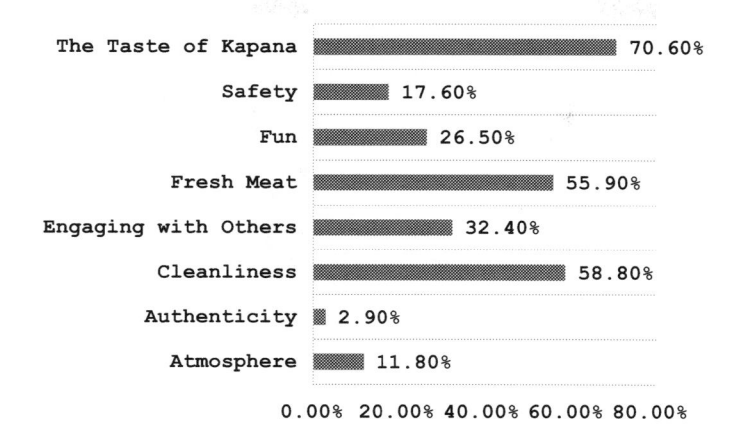

FIGURE 6.2 Elements influencing visitor experience (survey).

Engaging with others was referenced as the second highest in the online reviews, whereas cleanliness was the second highest in the empirical survey, as shown in Figure 6.2. Additionally, cleanliness and fresh meat ranked sixth highest in the reviews. Cleanliness is an essential aspect of the food service industry and one of the influencing factors on visitors' experience when visiting tourist attractions (Wiatrowski et al. 2011). Moreover, some food hygiene indicators, like freshness of food, food temperature and cooking level, are experience features and cannot be evaluated before purchasing (Yasami et al. 2020). Even though guests have always given value to cleanliness (Kumar et al. 2021), guests' concern for ethical hygiene has increased (Choi, 2019), so much so in light of the COVID-19 virus (Kumar et al. 2021; Syuhirdy et al. 2022). In such contexts, the existence of reliable and observable food safety indicators is a cue for dining and revisit decisions.

The authentic experience was referenced as the third highest in the online reviews, whereas the freshness of the meat was the third highest in the empirical survey. According to Băltescu (2016), providing authentic travel experiences is a prerequisite for increasing the attractiveness of a holiday destination. Tsai and Wang (2016) emphasised that the unique food image can be promoted in the competitive tourism market to distinguish a city from its competitors. Therefore, the development of authentic and unique street food by integrating the traditional menus and ingredients to highlight local authenticity and ensure the differentiation of the cuisine offered from its competitors.

Comparing the results from the online reviews, which mainly consisted of international visitors, and those of the empirical survey, which consisted of the locals, only shows that the level of influence of most factors on their experience differs significantly. For example, safety is ranked fourth highest among international visitors, whereas among locals, it is sixth in the ranking. Reisinger and Mavondo (2005) indicate that safety and security are essential factors that tourists consider when making travel decisions, especially when visiting overseas places unfamiliar to them.

Similarly, fun ranked higher among the locals than the international visitors. However, the atmosphere is among the least listed elements by international and local visitors. The atmosphere (ambient) conditions pertain to the intangible background environment (Ryu & Jang, 2006), including air quality, temperature, odour, music and sound. Since these may be perceived by the sense organs (Lin & Worthley, 2012), they can evoke sensory perceptions, influencing dining and revisiting decisions.

In light of the post-COVID market visit decision, the online reviews could not be used as reference because most were posted before COVID. However, the empirical survey revealed that all respondents were repeat visitors, with 67.6% having a frequency of one to two times and 32.4% having visited the market four to five times every six months prior to COVID, as shown in Figure 6.3.

Figure 6.4 shows that 76.5% indicated that their frequency of visits to the market has changed; 50% reported a decrease, whereas 26.5% indicated an increase in their visits to the market. The remaining 23.5% reported no change in their frequency of visits to the market.

Figure 6.5 shows that the factors which have influenced the visits to the market post-COVID are cleanliness and hygiene (61.8%), making it the highest influencing factor; the cost of Kapana is the second highest influencing factor; and safety was the least indicated influencing factor (23.5%). The above findings show that locals are more concerned about the cleanliness and hygiene of the street food market than the overall safety of the market.

Several elements were identified as critical in ensuring repeat visits to the Kapana market in the post-COVID era. Figure 6.6 shows that the taste of the

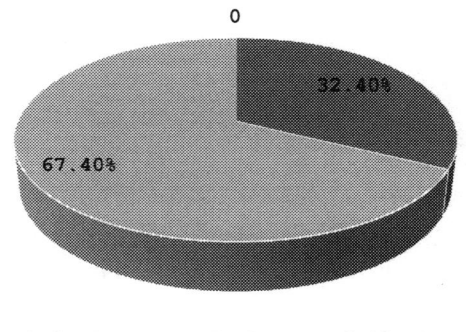

0

32.40%

67.40%

▪ 1-2 Times   ▪ 4-5 Times   ▪ 6-10 Times

FIGURE 6.3   Visit frequency.

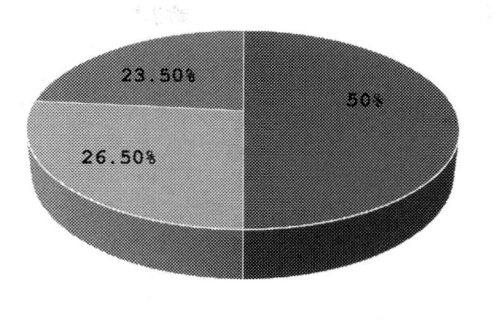

■Decreased ▨Increased ▨The Same

FIGURE 6.4   Visit frequency changes.

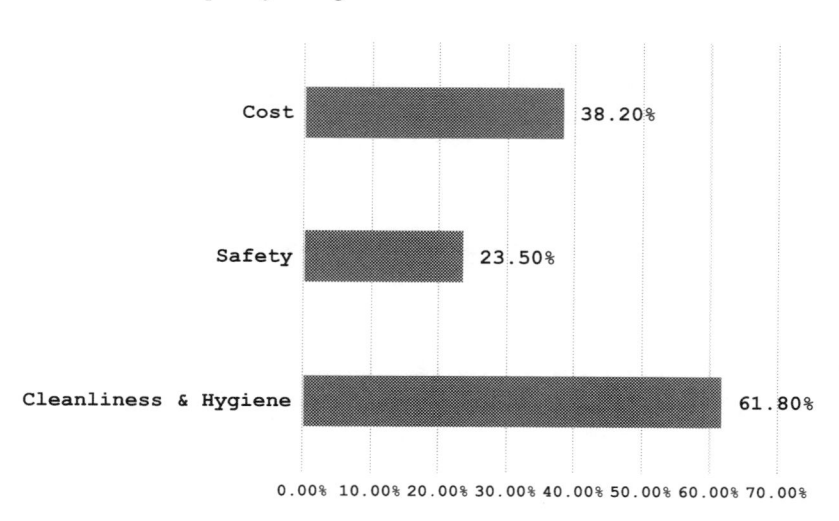

FIGURE 6.5   Post-COVID influencing factors.

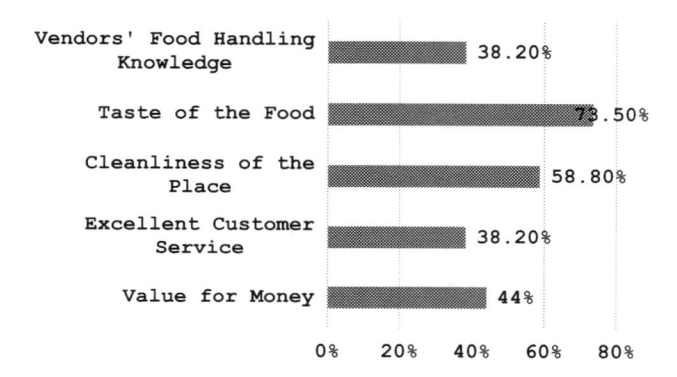

FIGURE 6.6   Re-visit elements.

food ranked the most important (73.5%), followed by the place's cleanliness (58.8%). The value for money was the third highest (44.1%). Excellent customer service and vendor's food handling knowledge was ranked 4th in importance (38.2%).

Conclusions

The study concludes that the influencing power of satisfying experiences is significant for consumer decision-making. Hence, the amalgam of satisfying factors can influence guests' revisit intentions. The taste of food and the cleanliness of the food market are competitive criteria for visitors' revisit intention and positive word of mouth, as seen in the survey that visitors who had good experiences are willing to recommend Kapana to others.

The value for money, or in other words, *the price*, is identified as one of the essential means of indicating quality and expected satisfaction (Ding et al. 2016; Živadinović, 2020), mainly because a higher price for a particular product or service indicates higher quality and vice versa (Kiatkawsin & Han, 2019).

Therefore, the Kapana vendors will do well to maximise the taste of meat that the visitors enjoy, as well as maintain a high level of hygiene and cleanliness to avoid cross-contamination while influencing guests' decisions to patronise and revisit the market. Furthermore, excellent customer service and vendor's food handling knowledge were the least significant in customer satisfaction. Nonetheless, excellent customer service and quality have been linked to creating sustainable competitive advantage in the service industry (Chu et al. 2012) and in the field of tourism (Lee et al., 2011), and food tourism is no exception. Similarly, because of the growing concerns about foodborne diseases (Linscott, 2011), safe food handling practices should not be ignored, more so with street food due to its high health risk association (FAO & WHO, 2022). The findings reinforced that visitor satisfaction significantly predicts revisit intentions to the Kapana market.

References

Ahrens & Ahrens. (2017, June 9). Namibian Cuisine – A fusion of delights. [Blog Post]. Retrieved from: https://www.myguidenamibia.com/travel-articles/namibian-cuisine ---a-fusion-of-delights

Arsal, I., Woosnam, K.M., Baldwin, E.D. & Backman, S.J. (2010). Residents as travel destination information providers: An online community perspective. *Journal of Travel Research*, *49*, 400–413. http://doi.org/10.1177/0047287509346856

Atikahambar, Y., Zainal, A., Rahayu, M., & Mokhtar, R. (2018). Quality of food and tourists' satisfaction of Penang delicacies in predicting tourists' revisit intention. *International Journal of Academic Research in Business and Social Sciences*, *8*(12), 1606–1618.

Aziz, S.A.A., & Dahan, H.M. (2013). Food handler's attitude towards safe food handling in school canteens. *Proc. Soc. and Beh. Sci.*, *105*, 220–228. https://doi.org/10.1016/j.sbspro.2013.11.023

Băltescu, B.A. (2016). Culinary experiences as a key tourism attraction. Case Study: Brașov County Codruța. *Bulletin of the Transilvania University of Brașov Series V: Economic Sciences*, *9*(58), 2.

Birch, D., & Memery, J. (2020). Tourists local food and the intention-behaviour gap. *Journal of Hospitality and Tourism Management*, *43*, 53–61.

Cho, H.S., Byun, B., & Shin, S. (2014). An examination of the relationship between rural tourists satisfaction, revisitation and information preferences: A Korean case study. *Sustainability*, *6*(9), 6293–6311. https://doi.org/10.3390/su6096293

Choi, J. (2019). Is cleanliness really a reason for consumers to revisit a hotel? *Journal of Environmental Health*, *82*(5), 16+. https://link.gale.com/apps/doc/A606675468/HRCA?u=anon~8550acd9&sid=googleScholar&xid=7974229d

Chu, P.Y., Lee, G.Y., & Chao, Y. (2012). Service quality, customer satisfaction, customer trust, and loyalty in an e-banking context. *Social Behavior and Personality: International Journal*, *40*(8), 1271–1283.

Cusack, I. (2000). African cuisines: Recipes for nation building? *Journal of African Cultural Studies*, *13*(2), 207–225.

Ding, H., Cao, L., Ren, Y., Kim-Kwang, R., & Shi, B. (2016). Reputation-based investment helps to optimize group behaviors in spatial lattice networks. *PLoS One*, *11*(9), e0162781.

Ekka, B.K. (2017). Food safety in street food in developing countries. *Indian Journal of Research - PARIPEX*, *6*(2), 217–219.

Everett, S., & Aitchison, C. (2008). The role of food tourism in sustaining regional identity: A case study of Cornwall, South West England. *Journal of Sustainable Tourism*, *16*(2), 150–167.

FAO & WHO (2022). Food safety is everyone's business in street food vending. Retrieved from https://www.fao.org/3/cc0037en/cc0037en.pdf

FAO. (2009). Good hygienic practices in the preparation and sale of street food in Africa. Tools for training. Retrieved from https://www.fao.org/3/a0740e/a0740e00.htm

Gemeda, A., Amenu, K., Girma, S., Grace, D., Srinivasan, R., Roothaert, R., & Knight-Jones, T.J.D. (2023). Knowledge, attitude and practice of tomato retailers towards hygiene and food safety in Harar and Dire Dawa, Ethiopia. *Food Control*, *145*, Article 109441. https://doi.org/10.1016/j. foodcont.2022.109441

Gunasekar, S., Das, P., Dixit, S.K., Mandal, S., & Mehta, S.R. (2021). Wine-experienscape and tourist satisfaction: Through the lens of online reviews. *Journal of Foodservice Business Research*. https://doi.org/10.1080/15378020.2021.2006039

Hendijani, B.R. (2016). Effect of food experience on tourist satisfaction: The case of Indonesia. *International Journal of Culture, Tourism and Hospitality Research*, *10*(3), 272–282. https://doi.org/10.1108/IJCTHR-04-2015-0030

Hiamey, S.E., & Hiamey, G.A. (2018). Street food consumption in a Ghanaian Metropolis: The concerns determining consumption and non-consumption. *Food Control*, *92*, 121–127. http://doi.org/10.1016/j.foodcont.2018.04.034

Imathiu, S. (2017). Street vended foods: Potential for improving food and nutrition security or a risk factor for foodborne diseases in developing countries? *Curr. Res. Nutr. Food Sci.*, *5*, 55–65.

Jayarman, K., Lin, S.K., Guat, C.L., & Ong, W.L. (2010). Does Malaysian tourism attract Singaporeans to revisit Malaysia? *Journal of Business and Policy Research*, *5*(2), 159–179.

Jeaheng, Y., Al-Ansi, A., Chua, B., Ngah, A.H., Ryu, H.B., Ariza-Montes, A., & Han, H. (2023). Influence of Thai street food quality, price, and involvement on traveler behavioral intention: Exploring cultural difference (Eastern versus Western). *Psychology Research and Behavior Management*, *16*, 223–240. https://doi.org/10.2147/PRBM.S371806

Jeaheng, Y., & Han, H. (2020). Thai street food in the fast growing global food tourism industry: preference and behaviors of food tourists. *Journal of Hospitality and Tourism Management*, *45*, 641–655.

Kapiris, K. (2012). *Food quality*. (Ed) InTech.

Khairuzzaman, M., Chowdhury, F.M., Zaman, S., Al Mamun, A., & Bari, M.L. (2014). Food safety challenges towards safe, healthy, and nutritious street foods in Bangladesh. *International Journal of Food Science*, 1–9. http://doi.org/10.1155/2014/483519

Khuluse, D.S., & Deen, A. (2020). Hygiene and safety practices of food vendors. *African Journal of Hospitality, Tourism and Leisure*, *9*(4), 597–611. https://doi.org/10.46222/ajhtl.19770720-39

Kiatkawsin, K., & Han, H. (2019). What drives customers' willingness to pay price premiums for luxury gastronomic experiences at Michelin-starred restaurants? *International Journal of Hospitality Management*, *82*, 209–219. https://doi.org/10.1016/j.ijhm.2019.04.024

Kozinets, R.V. (2010). *Netnography: Doing ethnographic research online*. Sage Publications.

Kumar, S., Ghosh, S., & Monda, B. (2021). Training-up during COVID-19 in hotel housekeeping operations in selected hotels of Kolkata. *Indian Journal of Hospitality Management*, *3*(1), 89–95.

Lee, S., Jeon, S., & Kim, D. (2011). The impact of tour quality and tourist satisfaction on tourist loyalty: The case of Chinese tourists in Korea. *Tourism Management*, *32*(5), 1115–1124.

Lin, I.Y., & Worthley, R. (2012). Servicescape moderation on personality traits, emotions, satisfaction, and behaviors. *International Journal of Hospitality Management*, *31*, 31–42. https://doi.org/10.1016/j.ijhm.2011.05.009

Linscott, A.J. (2011). Food-borne illnesses. *Clinical Microbiology Newsletter*, 33 (6), 41–45. https://doi.org/10.1016/j.clinmicnews.2011.02.004

Monney, I., Agyei, D., & Owusu, W. (2013). Hygienic practices among food vendors in educational institutions in Ghana: The case of Konongo. *Foods*, *2*(3), 282–294.

Noor, R. (2016). Microbiological quality of commonly consumed street foods in Bangladesh. *Nutr. Food Sci.*, *46*(1), 130–141.

Oladipo-Adekeye, O.T., & Tabit, F.T. (2021). The food safety knowledge of street food vendors and the sanitary compliance of their vending facilities, Johannesburg, South Africa. *Journal of Food Safety*, *41*, e12908.

Ozcelik, A., & Akova, O. (2021). The impact of street food experience on behavioural intention. *British Food Journal*, *123*(12), 4175–4193. https://doi.org/10.1108/BFJ-06-2020-0481

Reisinger, Y., & Mavondo, F. (2005). Travel anxiety and intentions to travel internationally: Implications of travel risk perception. *Journal of Travel Research*, *43*(3), 212–225.

Rosales, A.P., Linnemann, A.R., & Luning, P.A. (2023). Food safety knowledge, self-reported hygiene practices, and street food vendors' perceptions of current hygiene facilities and services – An Ecuadorean case. *Food Control, 144*,109377. https://doi.org/10.1016/j.foodcont. 2022.109377

Rozekhi, N. A., Hussin, S., Siddiqe, A., Rashid, P.D.A., & Salmi, N. S. (2016). The influence of food quality on customer satisfaction in fine dining restaurant: Case in Penang. *International Academic Research Journal of Business and Technology, 2*(2), 45–50.

Ryu, K., & Jang, S. C. (2006). Intention to experience local cuisine in a travel destination: The modified theory of reasoned action. *Journal of Hospitality and Tourism Research, 30*(4), 507–516.

Samapundo, S., Cam Thanh, T. N., Xhaferi, R., & Devlieghere, F. (2016). Food safety knowledge, attitudes and practices of street food vendors and consumers in Ho Chi Minh city, Vietnam. *Food Control, 70*, 79–89. http://doi.org/10.1016/j.foodcont. 2016.05.037

Samapundo, S., Climat, R., Xhaferi, R., & Devlieghere, F. (2015). Food safety knowledge, attitudes and practices of street food vendors and consumers in Port-au-Prince, Haiti. *Food Contr., 50*, 457–466.

Sousa, S., Albuquerque, G., Gelormini, M., Casal, S., Pinho, O., Damasceno, A., Moreira, P., Breda, J., Lunet, N, & P. Padrão, P. (2022). Nutritional content of the street food purchased in Chişinău, Moldova: opportunity for policy action. *Int. J. Gastron. Food Sci., 27*(2022), 100456. https://doi.org/10.1016/j.ijgfs.2021.100456

Syuhirdy M. N., Zakaria, N., Salim, A., Hussin, S., Mahat, F., & Rohiat, M. A. (2022). A Study on hygiene and cleanliness towards customer satisfaction: A focus on sunway lost world of Tambun hotel. *International Journal of Social Science Research, 4*(3), 238–252.

Tsai, C.T.S., & Wang, Y.C. (2016). Experiential value in branding food tourism. *Journal of Destination Management and Marketing, 6*(1), 55–65.

UNWTO. (2017). *Second global report on gastronomy tourism.* [online]. http://cf.cdn.unwto.org/sites/all/files/pdf/gastronomy_report_web.pdf [Accessed on August 2023].

Verma, R., Patel, M., Shikha, D., & Mishra, S. (2023). Assessment of food safety aspects and socioeconomic status among street food vendors in Lucknow city. *Journal of Agriculture and Food Research, 11.* https://doi.org/10.1016/j.jafr.2022.100469

Wiatrowski, M., Czarniecka-Skubina, E., & Trafiałek, J. (2011). Consumer eating behavior and opinions about the food safety of street food in Poland. *Nutrients, 13*(2). https://doi.org/10.3390/nu13020594

Yasami, M., Wongwattanakit, C., & Promphitak, K. T. (2020). International tourists' protection intentions to use food hygiene cues in the choice of destination local restaurants. *Geo Journal of Tourism and Geosites, 29*(2), 583–596. https://doi.org/10.30892/gtg.29216-491

Živadinović, B. (2020). Tourist satisfaction with quality of service, food, atmosphere, and value for money in restaurants of major cities of the Western Balkans. *European Journal of Applied Economics, 17*(2), 19–33. https://doi.org/10.5937/EJAE17-27360

7

MOTIVATIONS OF BACKPACKERS WHEN CHOOSING A TOURIST DESTINATION

Claudia Marcela Rodríguez, Sandra Rojas-Berrio and Oscar Robayo-Pinzon

Introduction

Tourism worldwide has had an outstanding performance during the last few years. Figures published by World Tourism Organization (UNWTO), from 2009 to 2018, evidence an important evolution in international tourist arrivals to the different continents. Africa, in 2009 received 46,100,000 international tourists, and nine years later a considerable growth was identified with the arrival of 68,400,000 tourists. On the other hand, the Middle East received 49,000,000 tourists during 2009 and in 2018 received 59,000,000 tourists. As for the Americas, in 2009, they received 141,000,000 tourists and in 2018 215,700,000, thus showing considerable growth. Asia and the Pacific, in 2009, received 183,600,000 international tourists and in 2018 347,700,000. Finally, Europe in 2009 received 472,600,000 international tourists and in 2018 - 716,100,000 (UNWTO, 2019).

According to the information presented above, a significant movement of travellers who moved from their country of origin to another country is identified. Globally, a 158% growth is identified between 2009 and 2018 (2008 – 892,300,000 tourists, 2018 – 1,407,400,000 tourists). '2018 marked the ninth consecutive year of sustained growth in international tourism. A total of 1.4 billion tourists traveled the world in 2018. Tourism generated US $ 1.7 trillion in global exports in 2018' (UNWTO, 2019).

For this reason, delving into different theories regarding motivations, we identify the authors Ryan and Deci (2000) who are cited as referents in different academic research documents. They expose the theory of Push and Pull motivations, referring to intrinsic motivations as inner psychological forces that an

DOI: 10.4324/9781032637778-8

individual possesses when traveling. Likewise, they expose extrinsic motivations as the external forces (attractions of a given place that impel to travel).

To make this measurement possible, this research focuses on backpacker tourists, who are considered a 'subculture' that has been evolving from the Middle Ages to the 21st-century Martín-Cabello (2014). Younger travellers constitute one of the fastest growing segments globally, and within these, backpackers show a dynamic of great interest to the tourism industry. A better understanding of the spatiotemporal behaviour of these travellers in terms of their preferences and motivations is a very relevant challenge today (Martins & da Costa, 2023).

According to the literature, the backpacker can be defined as a traveller who creates his own path freely and without limits, seeking extreme and diverse experiences, with a low budget, making use of an economical means of transportation. They also seek to expand their social circle by getting to know the local inhabitants. In conclusion, the backpacker can be described with an adventurous profile far from traditional tourism (O'Regan, 2022).

Thus, there is an interest in identifying what are the motivations of backpackers when selecting a tourist destination? Consequently, the general research objective is to identify the motivations of backpacker travellers when choosing a tourist destination, considering that there are two main types of intrinsic and extrinsic motivations.

It is pertinent to identify the motivations of backpackers so that in the future, other research, which will be able to reach the present study, can adopt them as a reference in strategies to reactivate tourism, considering the sanitary emergency declared by COVID-19 that was generated in 2020 (WHO, 2020). This situation temporarily and indefinitely halted tourism worldwide once most of the governments of the five continents opted for border closures and confinement as measures to prevent the spread of the outbreak. As a result, the behaviour of tourists changed surprisingly, as they were forced to suspend their leisure and relaxation activities. Recent studies mention that 'the recovery of tourism worldwide is estimated to take at least three years' (Bremner, 2020).

Literature review

From a psychological point of view, motivations are a set of conscious or unconscious stimuli that drive an action. Passed to the tourism sector, motivations can be considered as the cause that pushes the tourist to undertake a trip (Beltrán & Parra, 2017).

Human beings live with constant needs, some physiological and others psychological, which according to the Theory of Needs (Maslow, 1943), can be hierarchised in a pyramid as follows: level one: physiological or basic needs such as food, sleep, sex and breathing. Level two: Safety needs, feeling protected from any physical or psychological danger. Level three: Social needs, affection, friendship, love. Level four: Self-esteem needs, self-confidence, approval and social recognition among others, and finally level five: Self-realisation needs; these are related to independence, competence and full realisation.

However, Kotler et al. (2003) mention that a motivation arises from a need as soon as it reaches a sufficiently high level of intensity; people by nature try to satisfy the most important needs first (level 1, 2 and 3). On the other hand, De la Cierva (2006) presents in his work a classification of objective and subjective tourism dimensions: tourism motives can be particular or general; classified as follows: general (subjective), escape or escape, rest vacations, gregariousness, habit and particular (objective), for health, culture, sports, trade, religion, fun and change of climate.

De la Ballina (2017, p. 61), mentions that tourists relate their motivations closely with the destination; in this way they positively perceive the attributes of the destination that match their motivations. On the other hand, Lu Hsu et al. (2014) mention the theory of Ryan and Deci (2000), referring to the Push and Pull motivational factors scheme, mentioning that intrinsic motivations refer to the external forces that are considered when traveling, for example, prices, tourist destinations, sun and beach. Likewise, extrinsic motivations are psychological which individuals have when traveling, for example, evasion, rest, fun, self-excitement, leisure and socialisation, and in the case of backpackers, it is to travel around the world.

Similarly, Lu Hsu et al. (2014) mention through Crompton (1979) the theory of nine motivational factors, implicit in Ryan and Deci's (2000) Push and Pull theory, in which seven of them are Push: 'Escape from an environment, Exploration, Self-evaluation (self-actualisation), Relaxation, Prestige, Regression (performing childhood behaviours), Improving family relationships, social interaction (friendship)'. And two are Pull: 'Novelty (looking for something different) and Education and Culture'.

Likewise, Araújo and De Sevilha Gosling (2017) mention that tourism decisions influence motivations through psychological factors; they expose a classification of four motivations: physical (physical and mental health), entertainment, rest (pleasure and stress reduction) and psychological related to emotional development by sharing or expanding interpersonal relationships (family and friends).

Additionally, Días and Cassar (2005) refer to cultural motivations, mentioning the expansion of knowledge and personal evolution through different cultures, as well as history, gastronomy, music, etc. In addition, the prestige motivations which reach the individual to recognition, social status and increase the ego. On the other hand Araújo and De Sevilha Gosling (2017) mention that motivations significantly generate an influence on consumption, considering that although they are not the only variables, they do make part of the tourists' behaviour.

On the other hand, Rey (2017, p. 70) proposes the black box model, which consists of showing the external and internal stimuli received by the tourism consumer and his response through his purchasing behaviour. The author mentions that external factors are stimuli coming from the macro-environment (demographic, economic, sociocultural, environmental, etc.) and the external microenvironment (suppliers, intermediaries, customers and competition).

'For example, cultural factors include values, beliefs, preferences and behaviors learned throughout life' (De Borja Solé et al., 2002), conditioning how and where the consumer travels. Social factors refer to the influence of the social groups of the tourist consumer, such as family and friends. Personal factors refer to age, socioeconomic level, lifestyle, occupation, etc. Finally, the author refers to psychological factors such as the choice to purchase tourism products and services: 'it is also influenced by four factors'.

1. *Motivation*, defined as the amount of effort a tourist is willing to devote to satisfying his or her needs.
2. *Perception* refers to the way in which each tourist interprets the stimuli he/she receives from the outside.
3. *Learning* defines the level of knowledge of tourism products and destinations that the tourist has acquired through experience, and how it translates into behavioural changes.
4. *Personality*, which is related to the individuality of each tourist consumer and is specified in terms of the consumer's consumption and lifestyle.

Cooper et al. (2007) refer to the development of the theory of classifying the US population into a series of interrelated psychographic types:

1. *Psychocentric type*: this word is derived from 'psyche' and 'egocentric'. These are people whose behaviours are focused on small thoughts and concerns of small problems in life, their behaviour is egocentric, and they tend to visit safe places and return to them again and again, thus classified as 'regulars' (repeaters).
2. *The allocentric type*: this word is derived from 'alo' meaning 'varied form'. The behaviour of these individuals is adventurous, and they are motivated to discover new destinations, they rarely repeat places, hence the label 'globetrotter'.

Authors Cooper et al. (2007, p. 96) mention that most of the population focuses on these two profiles and call these extremes 'mesocentric'. Plog (1974) also mentions the close links between travel motivation and destination types. Allocentrics prefer unspoiled places to visit, while psychocentrics love the comfort of a safe and developed destination.

Finally, Araújo and De Sevilha Gosling (2017) relate in their research journal the TCL (Travel Career Ladder) theory known as travel career ladder, which is exposed by levels from the lowest to the highest, based on Maslow's pyramid.

As identified in the different elements of some of the documented studies mentioned above, it could be concluded that the authors coincide in proposing tourism motivations as intrinsic and extrinsic dimensions, which are born from a need to materialise when traveling. The study of tourism motivations is identified as a set of disciplines with different approaches that build a holistic view within the tourism sector. Table 7.1 shows some of the most relevant theories on motivations.

After analysing the literature, extensive research on tourism motivations was identified. However, the same result was not found regarding the motivations of backpacker tourism. Some scientific research papers mention that backpackers are defined as adventurers, who travel for long periods of time, organising their itinerary without expert advice, and generally traveling on a small budget. They are lovers of unpopular places, far from traditional tourism.

The following are some elements of documented studies on the subject.

In the 1980s, these tourists were identified as 'vagabonds' and were related to the hippie movement that came to be identified as 'backpacker' or 'backpacker', designing long duration trips with multiple destinations through a limited economic budget and looking for a deeper contact with nature.

On the other hand, Pérez and Asenjo (1999) in the preamble of their book *Manual del Mochilero*, narrate the origin of the word 'Mochilero', which comes from the Spanish word 'mocho', 'mochacho' and 'muchacho', derived over the centuries from the words 'mocho', 'mochacho' and 'muchacho'. From 'mutilu' we find in Basque 'mutil', 'young boy', 'mutila' adding the word 'a', later, 'mutxila' 'jovencillo', this was the name given in the north of Castile to the young boys who served and had the custom of going shopping at the market with baskets called 'mochilas'. Nowadays, 'backpacker' is a neologism that identifies the tourist who travels with a backpack on his shoulders, characterised by his own defined style of personally organising a trip and traveling the world at his own pace.

'The authentic backpacker is above all, a traveler who enjoys being in that new place, where he has arrived by his own initiative and ability, regardless of the conditions, efforts or hardships he has had to face' (Pérez & Asenjo, 1999). On the other hand, Revista Semana (2018) published an article in which reference is made to 'millennials', indicating that they changed large suitcases and first-class seats, for carry-on luggage and low-cost airlines. 'They seek experiences' (…) 'No suitcases, zero stress'.

Now, Tirado (2011), in his book *Traveling without Toilet Paper*, tells his story of adventure through different continents, mentioning that he travelled

TABLE 7.1 Theories of motivation, according to different authors

Author	Theory	Description of theory	Source
Ryan & Deci (2000)	Theory that suggests two schemes of motivational factors: Attraction and Push and Pull, called dimensions	**Push dimensions** refer to the inner psychological forces that an individual possesses when traveling. **Dimensions of attraction** refer to the external forces, attractions of a given place that drive people to travel.	Lu Hsu, J., Chun-Ting, T., & Yu-Hsin, P. (2014). Motivations for first-time and repeat backpackers in Shanghai
Crompton (1979)	Theory of nine motivational factors	**Seven nudges**: Escape from an environment, Exploration, Self-evaluation (self-actualisation), Relaxation, Prestige, Regression (performing childhood behaviours), Improving family relationships, social interaction (friendship) and Facilitation of social interaction. **Two of attraction**: Novelty (looking for something different), Education.	Lu Hsu, J., Chun-Ting, T., & Yu-Hsin, P. (2014). Motivations for first-time and repeat backpackers in Shanghai
Días & Cassar (2005)	Physical and Psychological Motivations	**Physical**: related to physical and mental health, entertainment, rest and stress reduction. **Psychological or interpersonal**: emotional development, visiting family and friends, establishing new relationships. **Cultural**: Personal evolution through other cultures and countries, breadth of artistic and historical knowledge. **Social or prestige**: they allow the individual to be recognised and appreciated.	Araújo, G., & De Sevilha Gosling, M. (2017). Los viajeros y sus motivaciones: un estudio exploratorio sobre quienes aman viajar

(Continued)

TABLE 7.1 (Continued)

Author	Theory	Description of theory	Source
Pearce (1988, 1993), Pearce and Caltabiano (1983) and Moscardo and Pearce (1986)	Theory of Travel Career Ladder (TCL) by levels from the lowest to the highest, based on Maslow's pyramid.	'The theory proposed that travelers advance up a hierarchy of travel motives as their travel experience increases'. However, tourists may change the order of their motivations over time. The theory describes five motivations: need for relaxation, need for security, need for relationships, need for self-esteem and development, need for actualisation and fulfilment.	Araújo, G., & De Sevilha Gosling, M. (2017). Los viajeros y sus motivaciones: un estudio exploratorio sobre quienes aman viajar. Paris, C. & Teye, V. (2010). Backpacker Motivations: A Travel Career Approach
Plog (1974)	Allocentric and Psychocentric Theory	**Allocentric:** travellers who seek adventure and experience new cultural and environmental challenges. **Psychocentric:** reputation and popularity.	Plog, S. C. (1974). Why destination areas rise and fall in popularity
Cierva (2006)	Tourism motives can be particular or general, a classification of tourism decision dynamics, objective tourism values and subjective tourism motives	**General (Subjective):** Avoidance, Holiday Enjoyment, Gregariousness, Habit. **Particular (Objective):** Amusement, Climate, Health, Culture, Sport, Professional or Occupational, Trade, Religion.	De la Cierva, R. (2006). Turismo, teoría, técnica, ambiente
Maslow (1943)	Hierarchical classification through a pyramid	1. Basic or physiological needs. 2. Security needs. 3. Affection needs. 4. Self-esteem needs. 5. Self-realisation needs.	Maslow, A. (1943). A theory of human motivation

Source: Own elaboration adapted to the studies cited above.

through atypical places on the planet and participated in controversial cultural situations as a 'backpacker', for him touring Europe taking pictures of monuments is a typical tourism that 'is extremely tedious'. He is a profile that perfectly describes an authentic backpacker, in his 329-page book he tells his experience, after studying in one of the best schools in Colombia and entering the University and doing different diplomas. He felt that in his life something was not working well, that is when he decided to take a trip with a simple backpack on his shoulders, without careful planning, looking for the most simplified way to enjoy his adventurous travels. Nowadays, backpacking tourism has spread to all continents and has been studied from different perspectives, including time (temporalities), geography (spatialities) and cultural norms (culturalities) (Timothy & Zhu, 2021).

In research conducted by Martín-Cabello (2014), he describes the concept of backpackers as a youth subculture created through networks in which the members, regardless of the origin of the country, generated a structure with their own meanings. Similarly, he alludes to the tourist and the traveller, mentioning through different authors that the tourist always plans his trip, while the backpacker does not know where to go. On the other hand, the tourist temporises the trip for weeks or months, while the traveller moves slowly from one place to another for years.

Ching et al. (2017) mention: 'Backpackers form an adventure tourism market are predominantly young people, which exploring the world on a tight budget, usually equipped with a backpack'. Through this research, different definitions of the profile of a backpacker traveller and their origin in Western nations are identified, with large participation in North America, Australia, New Zealand and Europe (Maoz, 2007).

On the other hand, Katrin and Hannam (2016) mention that independence is identified as a key element for backpackers in constructing trips, they also mention that 'Other studies that have looked at backpacking travel behavior suggest that independent tourists often seek unusual routes as well as crowd avoidance' (Elsrud, 2001; Muzaini, 2006; Sørensen, 2003).

Some backpackers are not comfortable with their familiar society and daily routine, so they look to travel to escape and seek opportunities for personal growth (Maoz, 2007).

Katrin and Hannam (2016) suggest a backpacker value system, describing it as a flexible 'code of honor' consisting of five criteria: (1) Travel with limited spending. (2) Meet new and different people. (3) Be free, independent, and open-minded. (4) Organise the trip autonomously. (5) Travel for as long as possible.

It is also important to mention that the predominant origin of backpackers is Western, mostly from North America, some from New Zealand, Europe and Australia; however, the population of Eastern backpackers has been growing in recent years (Maoz, 2007).

On the other hand, backpackers are not homogeneous groups, and although some stay in inexpensive places, it is estimated that they correspond to the middle and upper middle class, as they have expensive electronic devices such as photography equipment, tablets, laptops, high-end cell phones and credit cards. Likewise, some studies show that backpackers spend more than traditional tourists in the places visited (Martín-Cabello, 2014).

Regarding the motivations of backpackers, the literature is scarce; however, some studies have analysed results through various data collection methodologies. It has been identified that backpackers are motivated to travel for cultural knowledge, social growth, experience, independence and relaxation, among others (Paris & Teye, 2010). Another study reveals the possible influence that can exert motivation from sustainable consumption and have a lower impact on the destination when traveling as a backpacker (Agyeiwaah et al., 2021).

Lu Hsu et al. (2014) mention that regarding the Push motivational factors, Shanghai backpackers travel to improve skills, to do something specific, to challenge themselves, to increase knowledge, to meet new people, to see new places, to experience foreign lifestyles, to escape from daily life. On the other hand, regarding the pull motivational factors, some of them are availability of travel information, information about the city, ease of accessibility, high-quality restaurants, festivities, nightlife, historical sites, museums and art galleries.

According to the above, a backpacker is understood as a traveller who defines his own path freely and without limits, seeking extreme and diverse experiences, with a low budget, using a low-cost means of transportation and seeking to expand his social circle by meeting locals. It is identified as an adventurous profile, able to stay in hostels and places far away from traditional tourism, do not depend on tour guides, are identified by organising their itinerary without advice and independently, this allows them to accommodate their time and stay.

Principal objective

Identify the motivations of backpacker travellers when choosing a tourist destination.

Specific objectives

1. To identify the intrinsic motivational factors of backpacker travellers at the time of choosing a tourist destination.
2. To identify the extrinsic motivational factors of backpacker travellers at the time of choosing a tourist destination.
3. To describe the intrinsic and extrinsic motivational factors of backpacker travellers.

Methodological strategy

After conducting a thorough review of the literature, through secondary sources such as academic research papers with professional criteria, regarding the different theories of motivations, an important relationship is identified between two authors, Ryan and Deci (2000) with the self-determination theory and Crompton (1979), who exposes seven Push Motivations (which are a possible operationalisation of the intrinsic ones): 'Exploration and evaluation of the self, Escape from everyday environment, Prestige, Relaxation, Regression or nostalgia, Enhancing social relationships, and Facilitation of social interaction' and two Pull motivations (which are a possible operationalisation of the extrinsic ones): (Education and Culture, and Novelty). Thus, we proceeded to build an online questionnaire through Google Forms, which was adapted to the different elements of some documented studies mentioned above.

With the aforementioned information, a 57-question instrument was constructed. Survey completion was encouraged through two Amazon voucher drawings with a quota of US$50 for the first 200 participants and US$25 for the next 200 participants.

Thus, a target population was selected for this study, through different groups of the social network Facebook with profile of travellers Backpackers, these are: Viajeros-Mochileros, Viajeras por el mundo, Mochileros por Colombia, Mochiviajeros-Colombia, Viajeros por Colombia, Mochileros.org, additionally groups of travellers in Bogotá.

With the above, we sought to test the two hypotheses proposed for this study. H1: Motivations when choosing a tourist destination can be of a Push nature, which refer to inner psychological forces that an individual possesses when traveling, according to Ryan and Deci (2000). H2: Motivations when choosing a tourist destination can be of a Pull nature: 'external forces, attractions of a given place that impel to travel' according to Ryan and Deci (2000). For the design considerations of the instrument, a quantitative methodology was used, with descriptive levels of analysis and multivariate correlation.

After obtaining the results, the drawing for the Amazon vouchers was conducted with the support of the tutor assigned for the orientation of this research and recorded through the Google Meet application. Finally, the results were shared to the winners via email.

The mode of measurement of this study was based on the Likert scale method, with a rating of 1 to 5, being 1 never, 2 almost never, 3 sometimes, 4 almost always and 5 always. Thus, the target population of this study was selected through different groups of the social network Facebook with profile of Backpackers travellers, these are: Viajeros-Mochileros, Viajeras por el mundo, Mochileros por Colombia, Mochiviajeros-Colombia, Viajeros por Colombia, Mochileros.org. Additionally, the link to the survey was published on the walls of the profiles of these groups, and participation was encouraged by mentioning the Amazon voucher raffle. On the other hand, the completion of the survey was encouraged through a database with effective contacts that met the profile.

With the above, we sought to explore the two hypotheses put forward in this study:

> H1: Backpackers' motivations when choosing a tourist destination are Push in nature, which refer to the inner psychological forces that an individual possesses when traveling, according to Ryan and Deci (2000).

> H2: Backpackers' motivations when choosing a tourist destination are Pull in nature: 'external forces, attractions of a given place that impel to travel' according to Ryan and Deci (2000).

For the design considerations of the instrument, a quantitative methodology was used, with descriptive levels of analysis, with multivariate correlation.

In accordance with the objectives set out in this study, and in order to analyse the motivations of backpacker travellers when choosing a tourist destination, an exploratory quantitative analysis model was used through SPSS software, with the extraction method: principal components analysis and the Varimax rotation method with Kaiser normalisation, through a rotated components matrix, which analysed 44 elements. Likewise, the KMO sample adequacy measure and Bartlett's test were analysed with a result of .772 and a significance of 0.000, in addition to a percentage of 63.58% of the variance explained by the components of the extracted factors.

After obtaining the results, the Amazon vouchers were drawn and recorded through the Meet application. The drawing was conducted with the support of the tutor assigned for the orientation of this research. Finally, the results were shared to the winners via institutional email.

Findings

In order to highlight the most important demographics, it should be noted that, firstly, 71.8% of the respondents mentioned that they travel on a limited and/or reduced budget, 18.5% mentioned traveling on a large budget and 9.8% finance their trip. Likewise, 67% of the sample mentioned having a job, according to gender: 37% of the female gender and 30% of the male gender. In addition, when planning the trip, 50% of the sample asked for suggestions from other travellers, 23% sought advice from an expert, 14% arrived at the destination without planning the trip and 12% followed an influencer.

Secondly, 65% of the sample mentioned that their first trip was with their family, while 17% travelled with friends, 12% with their partner and 6%

travelled alone. Additionally, 93% of the respondents mentioned that their first trip was domestic and 7% mentioned that it was international. Finally, to the open-ended question, What was the last destination you visited? 100% of the respondents answered positively mentioning the destination.

Furthermore, the KMO master adequacy measure and Bartlett's test were analysed with a result of 772, as a result, the rotated components matrix yielded 10 factors, some of them with overlapping dimensions. Next, the results of the factor analysis will be presented in the order of element grouping, since, according to the first component, it is the one that reports the greatest total variance extracted and therefore is the one that manages to exhibit the greatest explanation (Hair et al., 1999) (Garza et al., 2013).

For a better understanding, Table 7.2 was prepared to show the results in a simplified form. Each of the factors is detailed below, starting with the results with reliability (>.702).

Pull motivations

The first factor yielded a reliability of .875, identifying a grouping of eight Pull factors overlapping between the patterns 'Education and Culture' and 'Novelty', through which it is confirmed that according to the literature (Crompton 1979), Pull motivations are understood as attractions of a given place that impel to travel. Thus, backpackers are motivated to learn and educate themselves through the culture of the countries visited. Variables such as history, gastronomy, cultural events, local lifestyles and economic alternatives in the market are factors that influence backpackers' motivations when choosing a tourist destination.

Push motivations

The second factor with a reliability of .702 grouped 6 items (5 of them from the construct 'improve social relations' and 1 of them 'relaxation' (I travel to rest and do nothing). With this behaviour it is identified that backpackers seek through travel rest and disconnection from the place of origin, enjoy the sun and the beach, be in natural environments or simply do nothing. Simultaneously, it is understood that through travel they seek to strengthen relationships with their friends, partners and family. In conclusion, backpackers seek to relax during the trip in the company of people close to their social environment, seeking to escape from their everyday environment.

The third factor has a reliability of .802. This factor identifies a grouping of six overlapping items in a combination of Push and Pull motivations (prestige, relaxation and novelty). This factor yields an important finding because when the mentioned dimensions are grouped in a natural way, a new motivation is

TABLE 7.2 Results

Factor	Cronbach's alpha	Cronbach's alpha based on standardised items	# items	Overlapping dimensions	Finding	Motivation
Factor 1	.875	.877	8	-Education and Culture (6) - Newness (2)	Education and Culture / Newness	Pull
Factor 2	.702	.702	6	-Relaxation (1)-Improving social relations (5)	Improving social relations	Push
Factor 3	.802	.804	6	-Prestige (2) -Relaxation (1) -Newness (3)	Entertainment	Pull & Push
Factor 4	.787	.785	5	Exploration and evaluation of the self	Exploration and evaluation of the self	Push
Factor 5	.784	.785	5	-Escape from the daily environment (4) -Regression or nostalgia (1)	Escape from everyday environment	Push
Factor 6	.908	.908	4	Facilitation of social interaction (4)	Facilitation of social interaction	Push
Factor 7	.862	.863	3	Regression (3)	Regression	Push
Factor 8	.518	.525	3	-Escape from everyday environment (1) -Prestige (2)	Prestige	Push
Factor 9	.540	.566	2	Relaxation (2)	Relaxation	Push
Factor 10	.493	.496	2	-Prestige (1) -Relaxation (1)	Prestige / relaxation	Push

Source: SPSS software – own elaboration.

identified which can be interpreted as an entertainment motivation; taking into account that backpackers mention being motivated by factors such as: night-time entertainment, fashionable places, sun and beach, carnivals and parties, and destination commerce, it is important to highlight that this motivation is not cited in Crompton's theory (1979), which is why it is qualified as a novelty in the present research.

The fourth factor, with a reliability of .787, confirms the push motivation 'exploration of the other self' by not overlapping any element; through this factor it is affirmed according to the literature, travellers feel the need to be alone for a while, to know themselves, to develop new skills, to make the trip a personal challenge and to explore independence. This finding is directly related to the 6% of the population that answered that they have travelled alone.

The fifth factor with a reliability of .784 grouped five dimensions (4 escape from the daily environment and 1 regression or nostalgia). Thus, traveling to get away from the work routine, escape from personal problems, calm the stress of everyday life and get away from the daily routine makes this dimension a need for social impact that in the future could be estimated in the new modality of ecotourism.

The sixth factor, with a reliability of .908 across four items, confirms the theory of (Crompton, 1979) regarding travel motivated by 'Facilitation of social interaction', alluding to using travel to make friends, interact with local people, meet people during the stay and have or improve social relationships. On the other hand, the seventh factor with a reliability of .862 confirms the push motivation 'regression', alluding that backpackers are motivated to travel to relive the past, remember the past and think about times lived.

Now, according to the results with reliability (<.541), three dimensions of Push nature of low impact were identified, these are:

The eighth factor with a reliability of .518 overlapped three elements; two of them corresponding to the motivations of 'prestige' (travel to tell others about my experience and travel to go to comfortable places) and one element of the motivation 'escape from everyday environment' (I travel to visit a recommended destination) understanding with this result that they are not strong dimensions in the decision of a tourist destination. It is understood that the backpackers surveyed are not motivated to brag about their trips or share their experiences with others, nor do they seek to visit recommended sites, understanding that they prefer to travel on their own, creating their own itinerary.

The ninth factor, with a reliability of .540, grouped two elements of the Push motivation, named by Crompton (1979) 'relaxation' (enjoying a natural environment and traveling for health). Finally, the tenth factor with a reliability of .493 yielded a grouping of two elements of the construct 'prestige' (I travel to show off trips) and 'relaxation' (practicing extreme sports), understanding that these motivations are not relevant for backpackers when choosing a tourist destination.

Regarding the last three factors described, it is concluded that traveling to show off their trips, traveling for health or to practice extreme sports, are Push motivations of low impact for backpackers at the time of choosing a tourist destination.

Discussion and conclusions

Taking as the basis of this research Ryan and Deci's (2000) Push and Pull theory, referring to Push motivations as the inner psychological forces that an individual possesses when traveling, and Pull dimensions as the external forces that attract to a particular place and drive to travel, and adapting Crompton (1979), the main purpose of this study was to identify the motivations of backpackers when choosing a tourist destination, and to discover if there is a relationship with respect to the two hypotheses raised.

Thus, through the factor analysis of the component's matrix, the factors of the Push motivations related to Crompton's theory (1979) were identified: exploration and evaluation of the other self, escape from the everyday environment, facilitation of social interaction, regression, prestige and relaxation. Some of these motivations take on greater force now of choosing a tourist destination, such as improving social relations, exploration and evaluation of the other self; escape from the everyday environment and facilitation of social interaction. On the other hand, it was identified that the Push motivations with the lowest denomination for backpackers are regression, relaxation and prestige. With this information, the first hypothesis (H1) is clear: The motivations of backpackers when choosing a tourist destination are of a Push nature, which refer to inner psychological forces that an individual possesses when traveling, according to Ryan and Deci (2000).

Regarding the second hypothesis (H2), the motivations of backpackers when choosing a tourist destination are of Pull nature (external forces, attractions of a particular place that drive to travel) according to Ryan and Deci (2000); the analysis of the components matrix was performed and once the first factor was identified through which eight dimensions of Pull nature were grouped, it is concluded that backpackers are motivated to choose the tourist destination attracted by Education and Culture, and Novelty, seeking intellectual enrichment during the trip, attending cultural events, enjoying fairs, carnivals and night-time activities and learning new things during the stay. They are also motivated by the destination's gastronomy and the market's economic alternatives. In this way, the second hypothesis (H2) is cleared, concluding that backpackers also choose the tourist destination for Pull motivations.

In view of the above, it is pertinent for the tourism sector to develop strategies aimed at the educational contribution of tourists, generating added value during their visit, understanding that backpackers seek to satisfy their need for intellectual formation through travel. To this end, value-added content can be implemented, through the history of the place and facilitation of visits to museums and historical sites, thus contributing to intellectual enrichment, offering knowledge of cultures and ways of life of the locals. Likewise, cultural events can be designed, offering the typical gastronomy of the region and thus generating economic offers in the backpacker tourism market.

On the other hand, it is important to highlight that in this research a significant finding was found in the factor explored in the rotated components matrix, through which three Push and Pull dimensions overlapped. It was clearly identified that backpackers choose the tourist destination to access night-time entertainment, attend fashionable places, enjoy carnivals and parties; they need the sun and the beach. Additionally, it was identified that tourists are attracted to the destination's commerce.

Now, with the natural behaviour of the grouping of these dimensions, a new finding is concluded: A motivation of Push nature that was called entertainment, understanding that backpackers seek fun and pleasure during their visit. Through this important finding, the tourism sector will be able to adopt strategies oriented to the satisfaction of its travellers in this sense.

Additionally, it should be mentioned that the motivation pushes 'facilitation of social interaction', which is one of the most relevant factors in the present study. It should be recalled that in March 2020 the World Health Organization declared COVID-19 as a 'pandemic' (WHO, 2020), from which worldwide containment was adopted as a prevention tool against this virus. Consequently, countries suspended commercial air and land transportation, and applied the measure of social distancing, which has had a psychological impact on society. Now, according to research conducted by several teachers from different universities in Spain, it is mentioned that the feeling of loneliness, pessimism and hopelessness presented a significant increase, 'This trend of change is greater when we consider those respondents who have symptoms or have been diagnosed of COVID-19, for those who are going through the confinement alone' (Lasa et al., 2020). Therefore, it is recommended that future research delve deeper into this factor to achieve a broader view on the motivation to travel as a facilitator of social interaction post-COVID-19, and thus determine its effect on the behavioural pattern when choosing a tourist destination.

On the other hand, travellers consider necessary a temporary change in their lives, choosing destinations that allow them to get out of that daily routine, seeking to 'Escape from the daily environment', this Push factor represents an important Insight as daily life and workload create the need to motivate themselves to disconnect for a while, as exposed by this factor. Thus, backpackers seek to get out of their work routine, escape from personal problems, relieve daily stress and, finally, eliminate depression. Additionally, backpackers choose destinations that allow them to 'explore the other self' and see travel as an opportunity to explore their identity. In this way, the tourist activity enables the development of new skills and becomes a personal challenge by allowing the individual to be alone for a while, get to know him/herself and explore his/her independence.

Likewise, it is perceived that backpackers choose a destination that allows them to 'regress' in time, remembering old times, and seeking to relive the past. In turn, it is identified in the results that the Push motivation 'To improve social relations' implies a determining factor at the time of choosing a tourist destination, considering that backpackers are motivated to strengthen family relationships, relationships with friends, with the couple, in addition to traveling to have fun with family and friends. This factor shows considerably that motivations are not necessarily tied to the destination, on the contrary their focus is based on people, and travel is a good opportunity to meet with family and friends to enrich their affective ties.

As a finding of the present study, three Push factors were identified as unreliable, which are considered of greater interest at the time of choosing the tourist destination:: 'traveling for prestige to show off the trips', 'traveling to tell others about the experience', 'traveling to be in comfortable places', and 'traveling to practice extreme sports' are not motivations that generate high impact for backpackers at the time of choosing the destination. Likewise, the Push dimension 'relaxation' shows that 'traveling for health' or 'to be in a natural environment' was not attractive to the target audience of this research.

Through the results of this research it was demonstrated that the most relevant motivations for backpacker travellers when choosing a destination are Push in nature: 'psychological inner forces that an individual possesses when traveling' (Ryan and Deci, 2000), identifying six factors that affirm this finding (see Table 7.3), Backpackers are motivated to choose the tourist destination motivated by the 'facilitation of social interaction', 'Escape from the everyday environment', 'exploration of the other self', 'regression' and 'improve social

TABLE 7.3 The 44 questions adapted to the push and pull theory

Motivation according to Ryan and Deci (2000)	Construct according to Crompton (1979)	Adaptation to questions
Push	Exploration and evaluation of the self	I am traveling to be alone for a while I travel to learn about myself I travel to develop new skills I travel as a personal challenge I travel to explore independence
Push	Escape from everyday environment	I travel to get away from work routine I travel to escape personal problems I travel to ease the stress of everyday life I travel to get away from my daily routine I travel to visit a recommended destination

(Continued)

TABLE 7.3 (Continued)

Motivation according to Ryan and Deci (2000)	Construct according to Crompton (1979)	Adaptation to questions
Push	Prestige	I travel to tell others about my travel experiences Nightlife entertainment (bars and nightclubs) I travel to go to trendy places I travel to go to comfortable places (with good hotels and/or restaurants) I travel to brag about my trips
Push	Relaxation	I travel to rest, to do nothing I travel to practice extreme sports I travel to enjoy a natural environment I travel for health I travel to enjoy the sun and beach
Push	Regression or nostalgia	I travel to learn about the history of the destination I travel to look back and think of the good times I have lived through I travel to experience nostalgia for memories I travel to relive the past or moments of the past I travel to get rid of depression
Push	Improve social relations	I travel to strengthen family relationships I travel to strengthen relationships with friends I travel to strengthen relationships with my partner I travel to have fun with family I travel to have fun with friends
Push	Facilitation of social interaction	I travel to make friends I travel to interact with local people I travel to meet new people while traveling I travel to have or improve my social relationships
Pull	Education and culture	I travel to enrich myself intellectually. I travel to learn about other cultures and ways of life. I travel to attend cultural events I travel to enhance my culture and education I travel to use travel to learn
Pull	Newness	I travel for the economic alternatives in the market I travel to enjoy gastronomy I travel for nightlife activities I travel to enjoy fairs and carnivals I travel for the commerce of the destination

Source: Own elaboration adapted from the cited authors Ryan and Deci (2000) and Crompton (1979) and scientific research documents.

relationships'; similarly, in the present work an important finding is identified which corresponds to the motivation to travel for 'Entertainment'.

It is important to mention that 78% of the population covered by this research mentioned having an average age between 20 and 39 years old, being the female gender the one that prevails in this age range. Therefore, it is estimated that women have a greater tendency to travel during their economically productive age, also 89% of the population mentioned having some type of higher education, which generates a correlation with the factor 'Education and culture', since travellers seek to enrich themselves intellectually during their trips.

Likewise, the results of the present work also indicate that Push and Pull motivations can be used as segmentation criteria for the tourism sector, understanding the need of backpacker tourists and adopting Push and Pull strategies to achieve a greater attraction to the destination, through emotions applied to training, social interaction, regression and encounter of the self, creating spaces for social interaction and offering entertainment.

In accordance with the experience of the present work, the theories of Ryan and Deci (2000) and Crompton (1979), which have allowed us to give an exploratory approach to the current research, achieving the objectives set, and identifying important findings in the population chosen for this purpose, should be taken as a reference for future research. However, the literature offers a wide variety of classification of motivations, through which it is also suggested to explore.

It is estimated for 2022 a post-COVID-19 global arrivals baseline in two scenarios: in the best-case scenario, a 50% reduction in global arrivals, and in the worst-case scenario, a 61% drop in the growth of arrivals (Bremner, 2020). For this reason, the motivations at the time of choosing a tourist destination will be of great relevance for the tourism sector and its corresponding reactivation worldwide.

Finally, it is important to mention that the implications for the development of the present work arose from the initial expectation of conducting the fieldwork with data collection in the Candelaria neighbourhood in Bogota, Colombia, through printed surveys to backpackers staying in hostels and hostels in this sector. However, due to the unexpected pandemic (COVID-19), the form of data collection was changed, using digital media for this purpose, achieving a timely and satisfactory result.

References

Agyeiwaah, E., Dayour, F., Otoo, F. E., & Goh, B. (2021). Understanding backpacker sustainable behavior using the tri-component attitude model. *Journal of Sustainable Tourism*, *29*(7), 1193–1214. https://doi.org/10.1080/09669582.2021.1875476

Araújo, G., & De Sevilha Gosling, M. (2017). Los viajeros y sus motivaciones: un estudio exploratorio sobre quienes aman viajar. *Estudios y perspectivas en turismo*, *26*(1), 62–85. https://www.redalyc.org/pdf/1807/180749182004.pdf

Beltrán, M. Á., & Parra, M. C. (2017). Perfiles turísticos en función de las motivaciones para viajar. *Cuadernos de Turismo, 39*, 41–65. http://doi.org/10.6018/turismo.39.290391

Bremner, C. (2020). Sustainability and digital transformation as recovery drivers TRAVE 2040. https://marketresearch.enterprise-ireland.com/wp-content/uploads/2019/01/Travel-2040-Sustainability-and-Digital-Transformation-as-Recovery-Drivers-Jul-2020-Euromonitor-.pdf

Ching, L., Suntikul, W., Agyeiwaah, E., & Tolkach, D. (2017). Backpackers in Hong Kong – motivations, preferences and contribution to sustainable tourism. *Journal of Travel & Tourism Marketing, 34*(8), 1058–1070. https://doi.org/10.1080/10548408.2016.1276008

Cooper, C., Fletcher, J., Fyall, A., Gilbert, D., & Wanhill, S. (2007). *El turismo: Teoría y práctica*. Editorial Síntesis.

Crompton, J. (1979). Motivations of pleasure vacations. *Annals of Tourism Research, 6*(4), 408–424. https://doi.org/10.1016/0160-7383(79)90004-5

De Borja Solé, L., Casanovas Pla, J. A., & Bosch Camprubí, R. (2002). *El consumidor turístico*. ESIC Editorial.

De la Ballina, F. (2017). *Marketing Turístico Aplicado*. ESIC Editorial.

De la Cierva, R. (2006). *Turismo, teoría, técnica, ambiente*. Editorial River.

Días, R. & Cassar, M. (2005). *Fundamentos do marketing turístico*. Pearson Prentice Hall.

Elsrud, T. (2001). Risk creation in traveling: backpacker adventure narration. *Annals of Tourism Research, 28*(3), 597–617. https://doi.org/10.1016/S0160-7383(00)00061-X

Garza, J., Morales, B., & González, B. (2013). *Análisis Estadístico Multivariante*. McGraw-Hill.

Hair, J. F. J., Anderson, R., Tatham, R., & Black, W. (1999). *Análisis Multivariante*. Prentice Hall International.

Katrin, A., & Hannam, K. (2016). The artisan backpacker: A development in Latin American backpacker tourism. *International Journal of Tourism Anthropology, 5*(1–2), 152–164. https://doi.org/10.1504/IJTA.2016.076853

Kotler, P., Bowen, J., & Makens, J. (2003). *Marketing para Turismo*. Pearson Education.

Lasa, B. N., Benito, G., Montesinos, H., Gorostiaga, M. A., Sánchez, E. J. P., García, P. J. L., & Germán, S. M. Á. (2020). *Las consecuencias psicológicas de la covid-19 y el confinamiento*. Servicio Editorial de la Universidad del País Vasco. https://www.euskadi.eus/gobierno-vasco/-/contenidos/documentacion/doc_sosa_consec_psique_covid19/es_def/index.shtml

Lu Hsu, J., Chun-Ting, T., & Yu-Hsin, P. (2014). Motivations for first-time and repeat backpackers in Shanghai. *Tourism Management Perspectives, 12*, 57–61. https://doi.org/10.1016/j.tmp.2014.08.001

Maoz, D. (2007). Backpackers' motivations the role of culture and nationality. *Annals of Tourism Research, 34*(1), 122–140. https://doi.org/10.1016/j.annals.2006.07.008

Martín-Cabello, A. (2014). El turismo «backpacker» en Chile como expresión de una subcultura juvenil global. *Cuadernos de turismo, 34*, 165–188. https://revistas.um.es/turismo/article/view/203071/164301

Martins, M. R., & da Costa, R. A. (2023). Motivations and spatiotemporal behaviour in an urban destination: A comparative analysis between backpackers from Generations Z and Y. In *Gen Z, Tourism, and Sustainable Consumption* (pp. 74–88). Routledge. https://doi.org/10.4324/9781003289586

Maslow, A. (1943). A theory of human motivation. *Psychological Review, 50*(4), 370–396. https://doi.org/10.1037/h0054346

Moscardo, G. M., & Pearce, P. L. (1986). Historical theme parks: an australian experience in authenticity. *Annals of Tourism Research, 13*(3), 467–794. https://doi.org/10.1016/0160-7383(86)90031-9

Muzaini, H. (2006). Backpacking southeast Asia. *Annals of Tourism Research, 33*(1), 144–161. https://doi.org/10.1016/j.annals.2005.07.004

O'Regan, M. (Ed.). (2022). *Backpacking culture and mobilities: Independent and nomadic travel.* Multilingual Matters.

Paris, C. & Teye, V. (2010). Backpacker motivations: A travel career approach. *Journal of Hospitality Marketing & Management, 19*(3), 244–259. https://doi.org/10.1080/19368621003591350

Pearce, P. L. (1988). *The Ulysses factor: evaluating visitors in tourist settings.* Springer-Verlag.

Pearce, P. L. (1993). Fundamentals of tourist motivation. In D. Pearce & R. Butler. (Ed.), *Tourism research: critiques and challenges* (pp. 85–105). Routledge and Kegan Paul.

Pearce, P. L., & Caltabiano, M. L. (1983). Inferring travel motivation from travelers' experiences. *Journal of Travel Research, 22*(2), 16–20. https://doi.org/10.1177/004728758302200203

Pérez, P., & Asenjo, I. (1999). *Manual del mochilero: aprender a viajar a tu aire.* Ediciones Desnivel S.L.

Plog, S. C. (1974). Why destination areas rise and fall in popularity. *Cornell Hotel and Restaurant Administration Quarterly, 14*(4), 55–58.

Revista Semana. (2018). *Bogotá principal destino turístico del país.* July edition.

Rey, M. (2017). *Marketing turístico: fundamentos y dirección.* Pirámide.

Ryan, R. M., & Deci, E. L. (2000). Self-determination theory and the facilitation of intrinsic motivation, social development, and well-being. *American Psychologist, 55*(1), 68–78. https://doi.org/10.1037/0003-066X.55.1.68

Sørensen, A. (2003). Backpacker ethnography. *Annals of Tourism Research, 30*(4), 847–867. https://doi.org/10.1016/S0160-7383(03)00063-X

Timothy, D. J., & Zhu, X. (2021). Backpacker tourist experiences: Temporal, spatial and cultural perspectives. In *Routledge handbook of the tourist experience* (pp. 249–261). Routledge. https://doi.org/10.4324/9781003219866

Tirado, D. (2011). *Viajando sin papel higiénico.* Medellín, Colombia.

World Health Organization. (2020). OMS coronavirus (COVID-19). https://www.who.int/es/emergencies/diseases/novel-coronavirus-2019/advice-for-public/q-a-coronaviruses

World Tourism Organization. (n.d.). Country profile – Inbound tourism. https://www.unwto.org/country-profile-inbound-tourism

World Tourism Organization [UNWTO]. (2019, January 21). *International tourist arrivals reach 1.4 billion two years ahead of forecasts.* https://www.unwto.org/global/press-release/2019-01-21/international-tourist-arrivals-reach-14-billion-two-years-ahead-forecasts

8

WHAT IS YOUR TRAVEL MOOD?

The effect of psychological flow on cultural *vs.* adventure travel experiences

Andrea Sestino and Cristian Rizzo

Introduction

The world of tourism experience is one of the most fascinating businesses, because of the ability to deliver to their clientele unique, extraordinary and memorable experience due to the cognitive and emotional individuals' involvement in experiencing such leisure activity (Bignè et al., 2008; Bigné & Decrop, 2019). Especially as the result of recent pandemic, individuals are increasingly calling for experiences able to give the opportunities to visit other places around the world (Zhong et al., 2021), both because of being forced in obliged lockdown periods and especially because of looking for new kind of hedonic and responsible experiences (Karasakal & Albayrak, 2022). Indeed, the total contribution of travel and tourism to the global gross domestic product (GDP) rose by 21.7% in 2022 over the previous years, after dropping sharply in 2020 due to the coronavirus (COVID-19) pandemic (Statista, 2022).

Reasons behind travelling basically consists in work, family, health or medical treatment, social or economic needs: Individuals who love to travel enjoy relaxation, new cultures, foreign food or incredible landscapes different to their normal surroundings (Huang & Hsu, 2009). Travelling may involve several cultural-related features in allowing individuals to catch new sights, learn new languages and learn about other cultures and different backgrounds (Plangmarn et al., 2012). Coherently, the cultural tourism consists in the movements of persons for essentially cultural motivations such as study tours, performing arts and cultural tours, travel to festivals and other cultural events, visits to sites and monuments, travel to study nature, folklore or art, and pilgrimages (Richards, 2013). Cultural tourists reach for situations able to emphasises

DOI: 10.4324/9781032637778-9

experiencing life within a foreign culture, rather than from the outside as a temporary visitor, by leaving their home environment at home, bringing only themselves and a desire to become part of the culture they visit (Jovicic, 2016; McKercher, 2020). Conversely, individuals also may travel because of being led by adventure motives, resulting in a type of tourism, involving exploration or travel with perceived risk, and potentially requiring specialised skills and physical exertion (Lindberg & Jensen; 2021; Millington et al., 2001). Adventure tourism gains much of its excitement by allowing the tourist to step outside their comfort zone. In this case, for example, tourists simply want to leave with a 'backpack' in search of mysterious, unknown places, engaging in adventures and situations far from their daily comfort zone in search of new and exciting stimuli (Huddart & Stott, 2020). Practically, adventure tourism may refer to explorations or journeys activities in which individuals may live in strict contact with the natural environment and, thus, require strong commitment, skills and a certain dose of physical efforts (Hudson, 2003; Ryan, 2013).

Individuals' reasons behind travel, literature demonstrate the central role of their involvement in tourism activities (Gursoy & Gavcar, 2003; Lu et al., 2015): Such a concept may also related to individuals' psychological flow consisting in a positive of intense engagement, involvement, focus and contentment in the present moment and current activity: It represents an individual tendency to feel a sense of 'staying in the flow', with no need to reflect, because the activity itself carries them forward as if by fairylike (Csikszentmihalyi, 2013). By considering individuals' psychological flow, literature demonstrate the positive effect of such a mental status in tourism activities, and specifically in promoting an overall satisfaction regarding the lived experiences (da Silva deMato et al., 2021; Karasakal & Albayrak, 2022): Indeed, the ability to promote a sense of flow may help tourism managers to create engaging experiences that will contribute not only to leisure activities but also to business activities, improving the tourists' chances of experiencing (p. 100818). Moreover, recent studies also add knowledge in this domain by demonstrating how the role of individuals' psychological flow may be also particularly effective both in adventure (Knowles, 2019; Lindberg & Jensen, 2021), and in cultural experiences (Jovicic, 2016; Li & Du, 2021).

Based on the above, in this chapter we explore the effect of a travel communication focus in tourism destination promotional campaign (adventure *vs.* culture) to investigate the effects on individuals' willingness to buy by revealing the effect of their psychological flow in leading such an effect. Results show that despite the travel communication focus does not directly influence individuals' willingness to buy a travel, the psychological flow may explain such an effect. More precisely, the adventure-based travel communication focus led to higher psychological flow and, in turn, higher willingness to buy; conversely, in the event of culture-based communication focus such effects decrease.

Theoretical framework

Essence of travel and individuals' travel motivation

Pine, Pine and Gilmore's (1999) view of the experience economy, intended to propose memorable experiences to the final consumers, totally reshapes the way to do business, especially in the tourism industry (Kim et al., 2012). Indeed, tourism research currents of thoughts focused on how to build unique tourism experiences able to guarantee an escape from everyday life, together with reaching strong effects on long-term memory (Quan & Wang, 2004; Larsen, 2007). Coherently, the issue of understanding the essence of certain tourism experiences represents a promising research issue regarding the experience of fruition of tourist destinations (Ritchie & Hudson, 2009; Richards, 2018).

As for individuals' travel motivations, seminal literature (Perce, 1982) explained how individuals may be attracted to some kind of the travel experiences because of the typicalities of the tourists' destination, because of their need to satisfy self-actualisation, love and belonging, and physiological needs: Indeed, tourists may find a tourism destination appealing not just for its scenic beauty, but also for the immersive and memorable experiences that make it a truly livable and enjoyable place to explore, and such opportunities may positively impact their reactions. Furthermore, by leveraging on Maslow's studies, literature (e.g., as in Pearce & Lee, 2005), proposed two important theories, namely travel career ladder, and travel career patterns, suggesting that individuals' travel needs may change over their life span and got travel experience, and thus, as tourists become more experienced, they increasingly seek satisfaction of higher-level needs.

In this scenario, the most relevant travel motives may include novelty-seeking, escaping from the routine and new relationship-seeking, because of being useful to positively impact individuals' seeking for different travel experiences related to their different life stages. Conversely, other travel motives consist of the nature of the travel (e.g., adventure, cultural, relaxation), self-development (host-site involvement), kinship, self-actualisation, self-enhancement, stimulation, isolation, nostalgia, autonomy, social status and romance (Oktadiana & Agarwal, 2002). In this regard, as anticipated, such kinds of travel experiences may include both adventure and culture travels.

Adventure travel represents one of the most fascinating kinds of tourism experience, and it is still partially under-investigated in academics (Beedie, 2012).

Adventure tourism is expressed in a type of tourist activity very often in close contact with nature, which requires physical, mental and endurance skills and abilities, capable of conferring sensations of excitement, finally allowing tourists to step outside their comfort zone, strongly going out from their ordinary daily life (Lindberg & Jensen, 2021; Rantala et al., 2018). Indeed, the nature-related experiences and the strict contact with the external environment is fundamental in influencing both individuals' motivations and experiences of adventure tourism activities, by focusing on its essence (Giddy & Webb, 2018). Such types of tourism experiences are generally performed in challenging situations in mountains, seas, lakes, deserts, far islands and so on useful – by their nature – to create unforeseen, exciting situations, useful to provide a pleasurable experience (Rojek & Urry, 1997). Indeed, one of the founding motivations of adventure tourism must be sought precisely in the adrenaline-pumping sensations and involvement (Giddy & Webb, 2018), and the maximum abstraction from reality (Weber, 2001) guaranteed by experiences of this type of tourism activities.

Conversely, cultural travels may be preferred because they are useful to stimulate cognitive experiences, and increase knowledge and personal culture, by facing daily life and situations typical of different cultures (Du Cros & McKercher, 2020; Jovicic, 2016): Such forms of tourism offer tourists the opportunity to 'experience first-hand' the cultural, social and historical heritage of a certain tourist destination on the basis of the conservation of cultural and artistic heritage, local prosperity for non-traditional tourist destinations, finally establishing links between different cultures as well (see Richards, 2018 for a review). Similarly, literature (Richards & van der Ark, 2013) also proposed that cultural tourists may develop a sort of cultural travel career, because younger visitors tend to consume more contemporary art, creativity and modern architecture, whereas older visitors are more prevalent at more traditional monuments and museums. By considering cultural tourism, the fundamental motivations behind such a travel approach are still widely investigated today (Richards, 2018), despite mainly resulting in cultural motivation, most clearly related to learning and knowledge (Richards, 1996), reaching for satisfaction and willingness to visit (Chang et al., 2014), and self-identity (Bond & Falk, 2013).

Interestingly, some worldwide destinations may offer – as a part of their tourism value proposition – both adventure and cultural travel, because of the typicality of their places. This is the case of India, which, with a perfect combination of culture and nature, may offer culture-related experiences (e.g., experiences related to spirituality) and adventure-related experiences (e.g., land-based, air-based and water-based comprising terrain vehicle tours, bungee jumping, safaris, as in the case of Uttarakhand mountains): A meaningful case in India is Rishikesh, known for its river rafting and cultural significance. River rafting in Rishikesh is one of the most popular adventure activities, offering the opportunity to raft down the River Ganga. Additionally, Rishikesh is referred to as the 'Gateway to the Himalayas' and the 'World Capital of Yoga' because of its diverse offerings. Visitors can experience adventure activities in the Himalayas as well as cultural and spiritual practices related to yoga (Chakraborty & Ghosal, 2022).

Similarly, Saudi Arabia offers both culture and adventure tourism experiences, as in the case of the Edge of the World in Riyadh that's known for its challenging cliffs and panoramic views to visit, and for the involving spiritual and mediation experiences related to the Muslim religion livable in the capital (Mumuni & Mansour, 2014; Rehman et al., 2023). Likewise, Europe, and specifically Iceland (Anna Om), Italy (Dolomites), Cappadocia (Turkey) and so on, may give tourists the opportunity to experiment with both adventure and culture travels.

Considering the peculiarities of both adventure tourism and cultural tourism, both seem to be able to stimulate positive sensations of total immersion in the activities to be carried out: on the one hand, adventure tourism can guarantee experiencing psychological flow because of the nature of the performed activities, as anticipated, closely related to natural environments and extremely challenging; On the other hand, cultural tourism could guarantee an experimental psychological flow as well, due to the total immersion that tourists could experience in 'getting confused' and touching cultural, historical realities far from their comfort zone.

Therefore, by extension, the focus of marketing travel communication relating to a tourism destination may be focalised on the livable adventure vs. culture contents, in the attempt to stimulate individuals' positive psychological flow, to ultimately impact their willingness to visit the promoted tourist destination.

The concept of psychological flow

The concept of psychological flow captured increasing attention from practitioners and researchers in the tourism field (e.g., as in Karasakal & Albayrak, 2022; da Silva deMato et al., 2021). The state of psychological flow indicates a mental state related to the feeling of total transport experienced by those who carry out an activity, able to abstract it from the real context in which it is located, making it totally focused on that activity (Csikszentmihalyi & Csikszentmihalyi, 1992; Csikszentmihalyi, 2013). Such a positive mental state typically derives from a positive individuals' mental attitude, and a high level of individuals' self-esteem (Asakawa, 2010). Recent literature in the tourism domain demonstrated the positive effects of the experienced psychological flow on individuals' overall satisfaction deriving from the tourism activities, and on their life satisfaction as well (Karasakal & Albayrak, 2022): Indeed, certain types of tourism activities (i.e., the adventure ones) can totally capture the attention of individuals, allowing them to immerse themselves in a situation of total transport such as to alienate them from the external environment.

Indeed, research has shown that both individuals' differences and motivations may positively influence individuals' seeking for flow-related sensations: More specifically, individuals' personalities and certain individuals' traits (Mills & Fullagar, 2008; Voiskounsky & Smyslova, 2003) may be considered as antecedents of individuals' proneness to experiment flow sensations, together with their motivations, because able to build certain skills useful to react to external/environmental stimuli to reach the flow activation (Ross & Keiser, 2014).

However, as aforementioned, the flow sensation may be also triggered by external stimuli in terms of the kind of activity to be performed. Indeed, the flow may be not only activated by variables such as the loss of cognition of time, the type of travel (i.e., adventure, and so on; Frochot et al., 2017), but also by situations in which individuals are engaged in activities in which a certain type of cognitive stimulation is required (Lee & Payne, 2016), as occurs, for example, in activities characterised by a high cultural content capable of abstracting bystanders from reality (e.g., as for observing a painting and getting lost among its colours; observing gestures and ways of doing of a local population, remaining totally involved and fascinated by them).

Thus, based on the reasoning below, we postulate that:

Hypothesis. *The effect of and adventure travel communication message focus (vs. culture) positively influence individuals' willingness to buy due to the effect of their perceivable psychological flow.*

The proposed conceptual model related to the proposed hypothesis is shown in Figure 8.1.

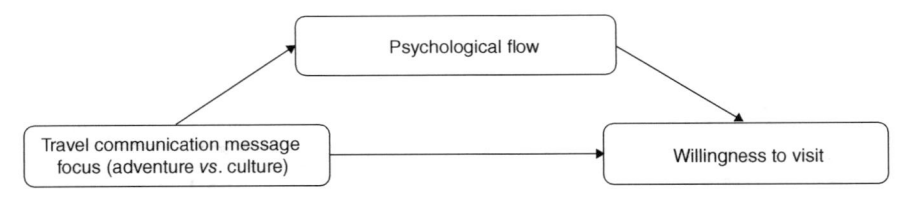

FIGURE 8.1   The proposed conceptual framework.

Methodology

Materials and sample

To test the hypothesis shown in the conceptual framework below, in this study, we manipulated the travel communication focus message (adventure *vs.* culture) two different scenarios.

In the adventure condition, participants read a scenario regarding an advertising communication message regarding a travel rich in adventure contents. In the cultural condition, participants read a scenario regarding an advertising communication message regarding a travel rich in cultural contents. For both the scenarios a fictitious travel destination has been chosen, by asking participants to imagine some places in Saudi Arabia.

To confer empirical support to our hypotheses, we conducted a survey among a sample of 1003 international randomly recruited participants aged between 18 and 79 years (M_{age} = 36.967, SD_{age} = 11.751), of which 59.621% males, and 40.379% females). Participants has been recruited trough Prolific software: By utilising such a tool, we also ensured that participants possessed a specific expertise, a passion for travel, and actively followed social network pages (e.g., Instagram) related to the specific travel type under study (adventure *vs.* cultural).

As for their education, the wide part of the final sample declared to hold a MSc (50.349%) or a BSc (37.787%), while the rest of them declared to hold a PhD (0.299%) or lower education levels (1.196% lower than high school; 10.369% high school or equivalent). Most of them declared to be employed (66.700%) or students (30.010%), while the rest of them declared to be unemployed (2.094%) or retired (1.196%). Finally, as for their geographical region, the collected final sample featured of 45.962% of American, 33.200% European, 19.940% Asian and 0.987% African respondents. The detailed sample analysis is reported in Table 8.1.

TABLE 8.1 Sample analysis (gender, education, employment, geographical region)

Sociodemographic	N	Percentage
Gender		
Male	598	59.621%
Female	405	40.379%
Education		
Lower than high school	12	1.196%
High school or equivalent	104	10.369%
BSc	379	37.787%
MSc	505	50.349%
PhD	3	0.299%
Employment		
Student	301	30.010%
Employed	669	66.700%
Unemployed	21	2.094%
Retired	12	1.196%
Geographical Region		
Northern America	351	34.995%
Southern America	110	10.967%
Europe	333	33.200%
Asia	200	19.940%
Africa	9	0.897%

Questionnaire

The administered questionnaire comprised three sections.

In the first one, participants have been welcome and assured about the anonymity of their response also in order to reduce their apprehension (Podsakoff et al., 2003); then, respondents have been randomly assigned to two different scenarios within a two-cell experiment that manipulated the communication focus of an advertised travel to buy (Adventure *vs.* Culture). In the second section, we then asked to rate their perceivable level of psychological flow by using an eight-items scale adapted from Jackson and Marsh (1996) and Ullén et al. (2012) (i.e., 'I have the clear will to reach the goal', 'I almost lose awareness of myself', 'I am in total control of the situation', 'I lose track of time passing', 'I feel that I am doing everything in the best way', 'I am very focused on what I do', 'I completely forget about personal problems', 'I feel totally taken by the situation and what I do'; $\alpha = .833$), and their willingness to buy the advertised travel by using the three-item scale drawn proposed by Dodds et al. (1991) (i.e., 'I would buy that advertised travel experience', 'I would consider to buy that advertised travel experience', 'The probability that I would buy that advertised travel experience is high', $\alpha = .769$). Both psychological flow and willingness to buy, were assessed on a seven-point Likert scale (1 = 'Strongly disagree', 7 = 'Strongly agree'). Finally, in the third section, we collected the main sociodemographic variables in terms of age, gender, education, employment and geographical location.

Results

To test our hypothesis, we ran the simple mediation model (Model 4) by Hayes (2013) of the PROCESS macro for SPSS (Table 8.2). The mediation model included the advertised travel communication focus (−1 = 'Adventure'; and 1 =

TABLE 8.2 Results of the statistical analysis

Dependent variable: Psychological flow (Me)	b	SE	t	p
Constant	5.368	.037	145.670	< .001
Communication focus (X)	−.093	.037	−2.531	< .001
$R^2 = .006$, $MSE = 1.212$, $F(1, 1001) = 6.404$ $p < .001$				
Dependent variable: Willingness to visit (Y)				
Constant	.503	.119	4.202	<.001
Communication focus (X)	.029	.025	1.160	<.001
Perceivable psychological flow (Me)	.899	.022	41.255	<.001
$R^2 = 0.631$, $MSE = .577$, $F(2, 1000) = 853.256$ $p < .001$				

'Culture') defined as the independent variable (X), consumers' willingness to buy the advertised travel as the dependent variable (Y), and their psychological flow which acted as a mediator (Me) in explaining the effect of the independent variable on the dependent variable. First, we regressed the psychological flow on the independent binary variable, and then we regressed consumers' intentions on the both the psychological flow, and on the independent variable, that is their willingness to visit. Results showed a significant and negative effect ($b = -.093$, $t = -2.531$, $p < .001$) of the type of communication focus on individuals' reaction. However, individuals' perceivable psychological flow exerted a positive and significant effect on their willingness to buy ($b = .899$, $t = 41.255$, $p = .000$). Moreover, results revealed a significant and direct effect of the type of communication focus ($b = .029$, $t = 1.160$, $p < .001$).

Thus, based on the above, the focus is not directly linked to consumers' reactions in terms of willingness to visit, but it is significant only among the effect of their psychological flow. More specifically, when the travel communication focus is on adventure, its effect on consumers' psychological flow increases and, in turn, their willingness to visit; Conversely, when the travel communication focus is culture-based, the effects on consumers' psychological flow and the total effect on their willingness to buy may decrease.

General discussion and conclusion

Tourism is one of the most important and promising industrial sectors nowadays both because it is able to contribute to hedonic consumption behaviours consistent with the experiential consumption paradigm, and because tourism activities are nowadays particularly sought after by individuals all over the world, who have experienced numerous lockdowns due to the pandemic, in isolation from their peers and prevented from traveling (Huang et al., 2022; Miao et al., 2022). Since tourists are now led by newer consumption motives more related to forms of responsible hedonic consumption (Erdoğan Tarakç & Yıldız, 2020; Karasakal & Albayrak, 2022), tourist experiences in close contact with nature or being able to contribute to the research of cultural contents are more appreciated now than in the past (Giorgi et al., 2021; Torres et al., 2022).

In this chapter, we investigated the effect of adventure vs. culture travel communication focus, on the potential tourists' willingness to visit a tourism destination, by shedding light on the role of individuals' psychological flow in leading such effect. Our reasoning derives from the relevance of psychological flow in tourism (da Silva deMatos et al., 2021) that may increase individuals' attitudes and behaviours towards certain tourism experiences because of the capability to stimulate mental and positive pleasure, and that may be exploited by tourism marketers and managers as an instrument to improve tourists' experiences (Karasakal & Albayrak, 2022).

Results from our experimental study demonstrated that despite the travel communication focus not directly affecting tourists' willingness to visit a tourism destination, the psychological flow may be a critical role in influencing such effect: Consequently, a travel communication focus empathising with the adventure features of a tourism destination may lead to a higher level of psychological flow and, in turn, of tourists' willingness to visit, rather than a travel communication focus emphasising cultural contents.

Theoretically, this chapter may contribute to tourism research and literature in several ways. Firstly, we add knowledge on the field of research related to the relevance of psychological flow in tourism (da Silva deMatos et al., 2021; Frochot et al. 2017), by revealing that despite being both adventure-related and culture-related, tourism activities features may potentially influence the flow sensations experienceable by the involved tourists, marketing stimuli mostly focused on adventure contents may be more effective in influencing such a status. Secondly, we add knowledge on the steam of research related to tourism destination management, by showing how in marketing and advertising campaigns aimed to promote a tourism destination, the focus on adventure *vs.* culture features of the promoted location may lead to different effects on potential tourists: By considering the peculiarities of the positive mental state associated to the sensation 'to stay in flow' (Csikszentmihalyi, 2013), tourism destinations marketers and managers should focus on adventurous experiences that their destination target of the marketing campaign may propose, rather than the cultural ones. Thirdly, our findings add knowledge in the domain of tourists' willingness to visit a tourism destination, by showing how the type of proposed activities may differently affect individuals' travel motivations and may have different effects on attracting them towards some kind of tourism travel and tourism destination (Pearce, 1982; Pearce & Lee, 2005).

Managerially, our results raise important considerations for tourism marketers and managers engaged in the design of marketing campaigns aimed at promoting the tourist destinations for which they strive, also in attempt to recover and ensure a rapid recovery of the tourism sector (Huang et al., 2022). Our results correctly answer the plea raised by da Silva deMatos et al. (2021), calling for research efforts in the domain of psychological flow applied to tourism. Moreover, we suggest some approaches, based on communication strategies, to catch potential tourists as a part of tourist destinations' efforts to sustain entrepreneurship in rural hospitality and tourism (Thirumalesh Madanaguli et al., 2021).

Based on our results, marketers and managers may leverage the feeling of psychological flow experienceable by the target tourists while visiting tourist destinations, to attract them and to incentivise their willingness to visit. Moreover, marketers and managers should carefully analyse the potential of their tourist destinations, to emphasise content related to adventure tourism *vs.* the contents related to cultural tourism, without, however, damaging the other reciprocally, but rather in an attempt to converge the proposed activities towards an 'all-encompassing' strong value proposition. Furthermore, findings of our research pose important reflections related to the concept of sustainable tourism, since they are focused on the conscious exploitation of natural and cultural resources of the targeted tourism destination. Adventure and cultural tourism activities required great exploitation of natural resources together with artistic, architectural and local heritage: In planning such stimulating tourism experiences, marketers and managers should be particularly attentive to prevent damage to these natural and cultural resources building a tourist destination; Indeed, they should work to better persevere over time these resources, making them enjoy as much as possible to the natives, and to the tourists who wish to visit and enjoy them.

Despite the promising and thought-provoking findings of our research, our chapter is not exempt from limitations. Firstly, the experiment has been conducted among a sample of randomly recruited international participants: Focusing on specific geographical locations or tourism destinations as a setting could reveal some international cultural differences influencing the overall results. Moreover, in this chapter, coherently with our research goal, we focused on the effect of the travel communication focus at the basis of a tourism destination advertising campaign on individuals' willingness to visit, by considering the role of their psychological flow: Some individuals' differences and sociodemographic variables (e.g., especially as for the role of age) have not been considered, if not in terms of covariates. Thus, future research could build on existing literature to explore the antecedents of psychological flow in adventure tourism, particularly among elderly consumers (Karasakal & Albayrak, 2022). Such research may also focus on different generational cohorts to highlight differences between various promising consumer segments, such as younger individuals, especially in the current post-pandemic world (Pop et al., 2022). Additionally, considering the hyper-technological landscape, future studies might examine the impact of new technologies on adventure tourism and how these innovations can help sustain cultural heritage (Massi & D'Angelo, 2020).

References

Beedie, P. (2012). Adventure tourism. In *Sport and adventure tourism* (pp. 228–265). Routledge.

Bigné, E., & Decrop, A. (2019). Paradoxes of postmodern tourists and innovation in tourism marketing. In Fayos-Solà, E., & Cooper, C. (eds.), *The Future of Tourism* (pp. 131–154). Springer.

Bigné, J. E., Mattila, A. S., & Andreu, L. (2008). The impact of experiential consumption cognitions and emotions on behavioral intentions. *Journal of Services Marketing*, *22*(4), 303–315.

Bond, N., & Falk, J. (2013). Tourism and identity-related motivations: Why am I here (and not there)?. *International Journal of Tourism Research*, *15*(5), 430–442.

Chang, L. L., Backman, K. F., & Huang, Y. C. (2014). Creative tourism: A preliminary examination of creative tourists' motivation, experience, perceived value and revisit intention. *International Journal of Culture, Tourism and Hospitality Research, 8(4)*, 401–419.

Chakraborty, P., & Ghosal, S. (2022). Status of mountain-tourism and research in the Indian Himalayan Region: A systematic review. *Asia-Pacific Journal of Regional Science, 6(3)*, 863–897.

Csikszentmihalyi, M. (2013). *Flow: The psychology of happiness*. Random House.

Csikszentmihalyi, M., & Csikszentmihalyi, I. S. (Eds.). (1992). *Optimal experience: Psychological studies of flow in consciousness*. Cambridge University Press.

da Silva de Matos, N. M., de Sa, E. S., & de Oliveira Duarte, P. A. (2021). A review and extension of the flow experience concept. Insights and directions for Tourism research. *Tourism Management Perspectives, 38*, 100802–100818.

Dodds, W. B., Monroe, K. B. and Grewal, D. (1991). Effects of price, brand, and store information on buyers' product evaluations. *Journal of Marketing Research, 28(3)*, 307–319.

Du Cros, H., & McKercher, B. (2020). *Cultural tourism*. Routledge.

Erdoğan Tarakçı, İ., & Yıldız, A. (2020). The effect of pandemic on consumer preferences in the context of hedonic and utilitarian consumption. *International Journal of Management, 11(7)*, 1527–1533.

Frochot, I., Elliot, S., & Kreziak, D. (2017). Digging deep into the experience–flow and immersion patterns in a mountain holiday. *International Journal of Culture, Tourism and Hospitality Research, 11(1)*, 81–91.

Giddy, J. K., & Webb, N. L. (2018). The influence of the environment on adventure tourism: From motivations to experiences. *Current Issues in Tourism, 21(18)*, 2124–2138.

Giorgi, E., Valderrey, F., & Montoya, M. A. (2021). Cultural tourism and the economic recovery of cities post COVID-19. In Montoya, M.A., Krstikj, A., Rehner, J., Lemus-Delgado, D. (eds) *COVID-19 and cities* (pp. 219–234). Springer.

Gursoy, D., & Gavcar, E. (2003). International leisure tourists' involvement profile. *Annals of tourism research, 30(4)*, 906–926.

Hayes, A. F. (2013). Mediation, moderation, and conditional process analysis. *Introduction to Mediation, Moderation, and Conditional Process Analysis: A Regression-based Approach, 1(6)*, 12–20.

Huang, S. S., & Hsu, C. H. (2009). Travel motivation: Linking theory to practice. *International Journal of Culture, Tourism and Hospitality Research, 3(4)*, 287–295).

Huddart, D., & Stott, T. (2020). What is adventure tourism? In *Adventure tourism* (pp. 1–9). Palgrave Macmillan. https://doi.org/10.1007/978-3-030-18623-4_1

Hudson, S. (2003). *Sport and adventure tourism*. Haworth Hospitality Press.

Jackson, S. A., & Marsh, H. W. (1996). Development and validation of a scale to measure optimal experience: The flow state scale. *Journal of Sport and Exercise Psychology, 18(1)*, 17–35.

Jovicic, D. (2016). Cultural tourism in the context of relations between mass and alternative tourism. *Current Issues in Tourism, 19(6)*, 605–612.

Karasakal, S., & Albayrak, T. (2022). How to create flow experience during travel: The role of destination attributes. *Journal of Vacation Marketing, 28(3)*, 303–318.

Kim, J. H., Ritchie, J. B., & McCormick, B. (2012). Development of a scale to measure memorable tourism experiences. *Journal of Travel Research, 51(1)*, 12–25.

Knowles, N. L. (2019). Targeting sustainable outcomes with adventure tourism: A political ecology approach. *Annals of Tourism Research, 79*, 102809–102811.

Huang, Y. T., Tzong-Ru, L., Goh, A. P., Kuo, J. H., Lin, W. Y., & Qiu, S. T. (2022). Post-COVID wellness tourism: providing personalized health check packages through online-to-Offline services. *Current Issues in Tourism*, 25(24), 3905–3912. https://doi.org/10.1080/13683500.2022.2042497

Larsen, S. (2007). Aspects of a psychology of the tourist experience. *Scandinavian Journal of Hospitality and Tourism*, 7(1), 7–18.

Lee, C., & Payne, L. L. (2016). Experiencing flow in different types of serious leisure in later life. *World Leisure Journal*, 58(3), 163–178.

Li, S., & Du, S. (2021). An empirical study on the coupling coordination relationship between cultural tourism industry competitiveness and tourism flow. *Sustainability*, 13(10), 5525.

Lindberg, F., & Jensen, Ø. (2021). Adventure regime of tourism experiences. *Current Issues in Tourism*, 24(20), 2905–2920.

Lu, L., Chi, C. G., & Liu, Y. (2015). Authenticity, involvement, and image: Evaluating tourist experiences at historic districts. *Tourism management*, 50, 85–96.

Massi, M., D'Angelo, A. (2020). *Reversing heritage destruction through digital technology: The Rekrei project*. In: Seychell, D., Dingli, A. (eds) *Rediscovering heritage through technology. Studies in computational intelligence* (vol 859). Springer.

McKercher, B. (2020). Cultural tourism market: A perspective paper. *Tourism Review*. 75(1), 126–129. https://doi.org/10.1108/TR-03-2019-0096

Miao, L., Im, J., So, K. K. F., & Cao, Y. (2022). Post-pandemic and post-traumatic tourism behavior. *Annals of Tourism Research*, 95, 103410.

Millington, K., Locke, T., & Locke, A. (2001). Adventure travel. *Travel & Tourism Analyst*, 4, 65–98.

Mills, M. J., & Fullagar, C. J. (2008). Motivation and flow: Toward an understanding of the dynamics of the relation in architecture students. *The Journal of Psychology*, 142(5), 533–556.

Mumuni, A. G., & Mansour, M. (2014). Activity-based segmentation of the outbound leisure tourism market of Saudi Arabia. *Journal of Vacation Marketing*, 20(3), 239–252.

Oktadiana, H., & Agarwal, M. (2002). Travel career pattern theory of motivation. In *Routledge handbook of social psychology of tourism* (pp. 76–86). Routledge.

Pearce, P. L. (1982). Perceived changes in holiday destinations. *Annals of Tourism Research*, 9(2), 145–164.

Pearce, P. L., & Lee, U. I. (2005). Developing the travel career approach to tourist motivation. *Journal of Travel Research*, 43(3), 226–237.

Pine, B. J., Pine, J., & Gilmore, J. H. (1999). *The experience economy: Work is theatre & every business a stage*. Harvard Business Press.

Plangmarn, A., Mujtaba, B. G., & Pirani, M. (2012). Cultural value and travel motivation of European tourists. *Journal of Applied Business Research (JABR)*, 28(6), 1295–1304.

Pop, N. A., Stăncioiu, F. A., Onişor, L. F., Baba, C. A., & Anysz, R. N. (2022). Exploring the attitude of youth towards adventure tourism as a driver for post-pandemic era tourism experiences. *Current Issues in Tourism*, 1–15.

Quan, S., & Wang, N. (2004). Towards a structural model of the tourist experience: An illustration from food experiences in tourism. *Tourism Management*, 25(3), 297–305.

Rantala, O., Rokenes, A., & Valkonen, J. (2018). Is adventure tourism a coherent concept? A review of research approaches on adventure tourism. *Annals of Leisure Research, 21*(5), 539–552.

Rehman, A. U., Abbas, M., Abbasi, F. A., & Khan, S. (2023). How tourist experience quality, perceived price reasonableness and regenerative tourism involvement influence tourist satisfaction: A study of Ha'il Region, Saudi Arabia. *Sustainability, 15*(2), 1340.

Richards, G. (2013). Cultural tourism. In *Routledge handbook of leisure studies* (pp. 505–514). Routledge.

Richards, G. (2018). Cultural tourism: A review of recent research and trends. *Journal of Hospitality and Tourism Management, 36*, 12–21.

Richards, G., & van der Ark, L. A. (2013). Dimensions of cultural consumption among tourists: Multiple correspondence analysis. *Tourism Management, 37*, 71–76.

Ritchie, J. B., & Hudson, S. (2009). Understanding and meeting the challenges of consumer/tourist experience research. *International Journal of Tourism Research, 11*(2), 111–126.

Rojek, C., & Urry, J. (1997). *Touring cultures: Transformations of travel and theory*. Routledge.

Ross, S. R., & Keiser, H. N. (2014). Autotelic personality through a five-factor lens: Individual differences in flow-propensity. *Personality and Individual Differences, 59*, 3–8.

Ryan, C. (2013). Risk acceptance in adventure tourism—Paradox and context. In *Managing tourist health and safety in the new millennium* (pp. 75–86). Routledge.

Statista (2022). Total contribution of travel and tourism to gross domestic product (GDP) worldwide from 2019 to 2021. https://www.statista.com/statistics/233223/travel-and-tourism--total-economic-contribution-worldwide/

Thirumalesh Madanaguli, A., Kaur, P., Bresciani, S., & Dhir, A. (2021). Entrepreneurship in rural hospitality and tourism. A systematic literature review of past achievements and future promises. *International Journal of Contemporary Hospitality Management, 33*(8), 2521–2558.

Torres, E. N., Wei, W., & Ridderstaat, J. (2022). The adventurous tourist amidst a pandemic: Effects of personality, attitudes, and affect. *Journal of Vacation Marketing, 28*(4), 424–438.

Ullén, F., de Manzano, Ö., Almeida, R., Magnusson, P. K., Pedersen, N. L., Nakamura, J., & Madison, G. (2012). Proneness for psychological flow in everyday life: Associations with personality and intelligence. *Personality and Individual Differences, 52*(2), 167–172.

Voiskounsky, A. E., & Smyslova, O. V. (2003). Flow-based model of computer hackers' motivation. *CyberPsychology & Behavior, 6*(2), 171–180.

Weber, K. (2001). Outdoor adventure tourism: A review of research approaches. *Annals of Tourism Research, 28*(2), 360–377.

Zhong, L., Sun, S., Law, R., & Li, X. (2021). Tourism crisis management: Evidence from COVID-19. *Current Issues in Tourism, 24*(19), 2671–2682.

9

PERCEPTIONS AND REACTIONS TOWARDS TIPPING FROM NON-TIPPING CULTURES

Employees' expectations

Eleanor Jayne Scanlon and Saloomeh Tabari

9.1 Introduction

Traditionally customers are expected to tip the employees for their service (Lynn, 2016; 2018). Tipping practices differ across cultures and countries; however, it is a global phenomenon (Lui, 2008) with it being a large aspect of many service industry workers. Tipping is considered a social norm within hospitality and in many service industries such as full-service restaurants, especially in the United States. Ferdman (2016) suggests that restaurant owners believe that tipping allows their clientele to express appreciation for good service. Tipping has a long history in the service industry and many studies have focused on this (e.g., Crespi, 1947; Shamir, 1984, Lynn and Graves, 1996; Lynn and McCall, 2000; Koku, 2005; Azar, 2005). The US-North American Industry Classification System (NAICS) stated that full-service restaurants are ;primarily engaging in providing food services to patrons who order and are served while seated and waiting for their meal and pay at the end of their meal' (U.S. Census Bureau, 2017). Customers or guests pay tips to employees (e.g., waiter/waitress, room service and porters) for waiting on their tables and assisting them with their dining/ staying experience. Tipping is not a new concept or trend in the United States and can be traced back to the 1800s (Segrave, 2009). Tipping is not only a norm in the United States; as in many other countries like Australia, Turkey, the United Kingdom, and Iran it is optional, a voluntary gesture and yet people still perform it. On the other hand, in some countries, Japan, for example, tipping conveys the opposite message of gratitude, meaning that the service was not considered good enough.

Tipping is becoming increasingly relied on from an employee perspective in the service industry due to low wages. According to the US Department of

DOI: 10.4324/9781032637778-10

Labour (2017), employees are paid a basic hourly wage by their employer that is less than the federal minimum wage because tips are considered their main source of income. Customers are aware of the importance of tipping the employees and consider it a norm. The fundamental logic of economic thoughts on tipping is that it is the most efficient way of observing and rewarding service employees who get paid a minimum wage (Bodvarsson and Gibson, 1994; Hemenway, 1984). A study conducted by Azar (2005) on the characteristics of tipped and non-tipped occupations shows a negative correlation between employees' income and consumers' rewarding ability. These findings are not only interesting but are also in disagreement with the prevailing economic view on tipping. If tipping means voluntarily paying more for a service than a customer is expected to, then the practice could be irrational given the view that consumers generally want to pay less for more (Mankiw, 2007; Landsburg, 1993). This is despite the fact that tipping could benefit the recipients, suggesting that organisations in the services industry that permit tipping could be taking on risks (Lynn and Wang, 2013).

Some of the common problems faced by service business organisations such as restaurants that permit tipping are optional behaviours by employees (Barkan et al., 2004). These behaviours include reduced efforts on the part of waiters where tips are shared equally among all employees, inclusive of the kitchen staff. It also leads to overuse of resources where individual employees keep their tips. As such, employees may frequently refill drinks, and bartenders could be generous in serving drinks at the expense of the employers to win customers' favours for a higher tip (Lynn and Graves, 1996).

Previous studies have focused on different areas of tipping, for example, Lynn and Latane (1984) investigated the effects of such factors as dining group size, customers' gender, method of payment and bill size. Cunningham (1979) and Koku (2005) examined the role of race and gender of tippers, whereas Bodvarsson and Gibson (1994) and Azar (2004) examined the economic underpinnings of tipping, and Azar (2007) examined the role of norms in tipping. Karabas et al. (2020) explored customer response to tipping requests at limited-service restaurants. Only a handful of research has focused on employees' expectations and motivation towards tipping. For example, Curtis et al. (2009) focused on Employee Motivation within tipped and non-tipped restaurants; Namasivayam and Upneja (2007) examined employee preferences for tipping systems. Lin and Namsivayam (2011), in one of the rare studies of its kind, examined employees' perception of tipping. Specifically, there is limited literature focusing on employees' expectations, motivation and attitude towards tipping, especially if they are from a non-tipping culture.

Previous studies examined the motivations of tipping, and the racial and ethical differences between tipping and consumer perspectives, but there is limited research on employee's expectations and the effects of tipping on their working ethic. Abraham (2014) states that most of the work comparing tipping differences is between groups focused on ethnic groups within the United States and that there is a need for more international research on the tipping conversation. This chapter aims to explore the effects of tips on employees from a non-tipping culture when working in a tipping culture like the United States of America. There is a gap in looking at employee's expectations and reactions when entering a culture where tipping is the norm. As a result, consider this study as exploratory research focusing on a sample of ten employees from a non-tipping culture and their expectations and reactions towards tips and ten employees from a tipping culture. The chapter research questions are:

- What are the impacts of tipping on employees' expectations, motivations and emotions?
- How does receiving and not receiving tips affect employees from non-tipping cultures?

This research contributed to the tipping culture and management strategy within the service sector and especially could work for small and independent services. Although, in some cases, tips can be considered a good culture for the service industry, there are still disadvantages of the tipping system and has been seen mostly as a bad culture that may lead to conflict, lack of teamwork, unhealthy competition and demotivation among employees. This will provide an opportunity for managers in the service sector to provide a solution to the tipping system and change the direction towards better practice within the organisation.

9.2 Tipping and its meaning

A significant part of the literature on tipping suggests that there is an emotional component to tipping (Azar, 2005). Matilla and Enz (2002: 268), in their study of frontline hotel employees, stated that 'consumers' evaluation of service encounters correlates highly with their displayed emotions during the interaction and post encounter mood states'. Tipping is now a means of a common business practice. It leads back to as far as the middle ages when lords would tip beggars for safe passage (Koku, 2005); in the 18th century tipping was introduced to coffee shops to ensure promptness of service, where patrons would drop a tip in a collection box (Lynn and Latane, 1984). Tipping has always been a practice of the service industry and continues to be.

Generally, there are several reasons for tipping, which include social approval, equitable service exchange (Saayman and Saayman, 2015), the value of the meal (Koku and Savas, 2016) as well as the interpersonal connection between the server and the customer as interpersonal similarity, in respect of the food service experience provided (Parrett, 2011). In this vein, the research by Gössling et al. (2020) indicates three ways of basic pricing systems in restaurants. The first is where a customer pays for the food itself, plus a gratuity of their choice, this would then make up the servers' income. The second way is where a service charge is added in addition to the menu price, and this is usually highlighted on the menu. The third way is service inclusive pricing where a service charge would already be added to the menu price.

There are cultures in the world like the United States and Egypt where tipping is seen as a necessary action to conform to; however, in places like the United Kingdom, New Zealand and Denmark tipping is less of a common conformity (Whaley et al., 2014; Lynn, 2000). There is confusion as to whether or not to tip for a certain service, people from global destinations such as Las Vegas, where there is a high percentage of hospitality staff, will often rely on having successful intercultural encounters, but first, they must overcome such issues as language barriers and cultural biases (Abraham, 2014) as well as the global confusion of tipping. Rokou (2014) identifies that Americans are amongst the most gratuitous when it comes to tipping, having found that 56% of a TripAdvisor survey think that tipping is expected and one-third feel obligated to tip even when the service is poor. Swedish website Travel Forum provides an overview of service gratuity expectations in 250 countries and regions, suggesting that it is uncommon to tip in Belgium or Denmark, and customary to tip up to 20% in Canada or the United States.

Lynn (2000) identifies that tipping is more likely to be prevalent in cultures with extroverted populations rather than introverted populations due to the idea that the value of things, such as social functions, like tipping is higher. Social norms are a high motivator for whether or not a person will tip, and the prevalence of a tip and its size will differ along with national attitudes. In support of the last research, Lynn and Starbuck (2015) mention that a national extraversion will have a positive effect on tipping. The characteristics above all indicate societies in which social functions are valued more. In contrast, Saayman and Saayman (2015) point out that people from South Africa generally tip when their service has been good rather than a status symbol showing off their income.

9.3 Employee's expectations and motivation

Many workers who are in a tipped position, for instance waiters, rely on their tips as a significant portion of their income. So, when a tip becomes inadequate due to reasons such as being uneducated about a tip, unsure of what and when to tip and possible demographic differences it can create a negative impact on the server, restaurant and management team (Abraham, 2014). It is identified that in the United States alone US$47 billion a year is made in tipping in the food industry (Azar, 2011). Saunders (2015) mentions in his research that a tip gap approach was identified. He states the effects of the difference between the perceived/expected tip and the actual tip received. Identifying that the employee's effective state will become negative the lower the tip.

From a managerial perspective, motivation and tips can significantly affect the performance of a server (Whaley et al., 2014). Thus, resulting in when employees of the industry go to dine out themselves, they on average tip 4% to 5% more than customers without experience in the food industry (Parrett 2011). Several managerial implications have led to recognising the importance of tipping expectations and the influence they play on the displayed emotions of the servers in the workplace (Saunders, 2015). An employee's tip expectations are greater than what most other social etiquettes would recommend, this being because managers exaggerate tip expectations to attract and retain staff as well as the employee competition seen in the workplace about who can make the most money (Dublanica, 2010).

Shamir (1983) identifies that the more closely monetary rewards are related to job performance, the larger the expected worker motivation. The tip is a very immediate monetary reward and therefore creates a higher sense of motivation than a normal wage, indicating that the workers who receive tips are more motivated than those who do not receive tips. Furthermore, Curtis et al. (2009) believed that the top motivational factors of an employee in the restaurant industry are management loyalty, good working conditions, job security and good wages. They also stated that the motivation scores between non-tipped and tipped employees were different for many reasons. Those of a non-tipped employee would see more job progression, manager recognition and promotion. Seiter and Weger (2010), in their research, examined the tipping differences in party sizes and how this would affect the size of a tip. In their study the servers were asked to complement all orders. It was confirmed that a party size of five or more diners has a negative effect on the efficiency of generalised compliments coming from the server and therefore negatively impacting tipping behaviour.

9.4 Race and ethical differences

A research based in the United States identified that race does make a difference when it comes to tipping. McCall and Lynn (2009), in their survey from servers, indicate that 65% rated African Americans as below-average tippers. This makes restaurant managers struggle when it comes to retaining and hiring staff in predominantly black neighbourhoods (Lynn, 2011; Amer, 2002). Previous research suggested that in Black neighbourhoods they are less educated in the sense of tipping and the average American tip of 15–20% is not as commonly known. In agreement with this Noll and Arnold (2004) state that on average black parties would almost always tip below 15% in comparison to white parties who would on average tip above 15%. In different research by Lynn and Brewster (2015), it has been identified that the mediators between a tipping norm, race and ethical differences are sex, age, education, income, experience working for tips and counterbalanced order. However, it is important to understand that white people tend to receive a better quality of service as compared to blacks (Were, Miricho and Maranga, 2019).

It has been suggested that servers who perceive low levels of workplace autonomy or low-wage security will receive smaller tips from their black customers compared with their white customers. The relationship will be mediated by the amount and degree to which servers withhold or extend subtle behaviours such as smiling, entertaining and touching, during the service interaction. Servers go to less effort towards certain people, such as 'rednecks', women and seniors, who, by servers, are perceived to be below-average tippers by withholding subtle behaviours such as smiling and touching, which are common actions associated with male customers (Brewster and Mallinson, 2009). Koku (2005) mentions that the relationship between race and tipping is well known in the respect that minority group members are not good tippers. This generalisation leads to a form of stereotyping on servers whether they consciously or unconsciously realise that they believe they will be poorly tipped by a minority patron. This coincides with the minority patron, who feels they are less well treated, leading to a result of them being less likely to leave a tip.

Lynn (2011) suggests that if you were to eliminate race differences in the awareness of a tip rate of 15 to 20%, the restaurant tipping norm would reduce race differences in restaurant tip percentages by only about 30%. Thus, indicating that even if the black awareness of the tipping norm is brought up to the level of whites, the race difference in tipping will exist. In addition, Lynn and Brewster (2015) introduced a model of relationships and their expected effects on tip size has been created, using variables which have come from an online survey of a diverse sample of Asians, blacks, Hispanics and Whites.

9.5 Customer's perspective

Whether or not a tip is expected, it is proven in a study by Koku and Savas (2016) that there is a positive effect that a person's emotional contagion will be affected by the waiter's attitude (Body language and eye contact), service and experience. A customer at times can use a tip to signify whether or not they enjoyed their experience without having to personally confront the server (Whaley et al., 2014). Moreover, Margalioth, Sapriti and Coloma (2010) stated that people do not want to risk social disapproval, resulting in customers opting to fulfil the norm of tipping. However, the tip can be affected by the amount of verbal contact that the employee has with a customer. Employees' verbal attention does affect the customers' buying intentions and tipping behaviour. It was found that more dining groups made an extra purchase and/or left a tip when the employee paid particular verbal attention to them (Ivkov et al., 2017).

Looking at the customer from a different perspective, customers with experience as a server will in general tip higher than a customer without. Customers again with experience as a server have a tendency to notify their server that they have the experience in the hope that this will increase their satisfaction with the meal experience (Parrett, 2011). Another factor which can affect the size of a tip from a customer is their income and occupation; some people with a high income perceive their ability to pay a personal satisfaction (Saayman and Saayman, 2015), thus leading to tipping on a more frequent basis. A tip is what you pay to show thanks for the service you receive. When tipping is taken away from the equation, customers will now expect to see higher menu prices and the insufficiency to reward their servers (Koetke, 2016). However, tipping carries on being a voluntary act of the individual customer; therefore, the reasoning behind a tip lies in the individual human motivation.

Although a range of previous literature on tipping – behavioural, experimental economics literature – as well as theory and research, and other social science disciplines have been analysed to create this framework, Lynn (2015) selectively drew from the literature of the primarily driven motivations. It is stated that although there are racial, ethical and social processes which affect a tip, a person's motives are still the main motivation.

9.6 Methodology

The purpose of this research is to consider employees' attitudes towards tipping when they are not from a tipping culture. A qualitative approach has been employed, based on the following questions:

- What are the impacts of tipping on employees' expectations, motivations and emotions?
- How does receiving and not receiving tips affect employees from non-tipping cultures?

As mentioned earlier, only a handful of studies focus on employee expectations towards tips (See Saayman and Saayman, 2015; Parrett, 2011; Lynn and Simons, 2000; Lynn and Brewster, 2015), and none of the existing research focuses on those employees who are not from tipping cultures but are working in or desire to work in, a tipping environment. Semi-structured interviews using purposive sampling were undertaken in the United Kingdom. In total 20 interviews were carried out, all involving participants from non-tipping cultures who have a degree in hospitality management. The purposive sampling technique was adopted, as it allows researchers to select informants who are familiar with particular settings, persons or groups. They are intentionally selected for the important information they can provide, which cannot be developed as successfully from other choices to find a specific answer (Creswell and Plano Clark, 2010; Neuman, 2017). For the purpose of this chapter, we have chosen two groups of participants: one group who have worked in a tipping culture environment (10 participants) and a second group who desire to work in a tipping culture environment (10 participants). Each interview lasted between 30 and 40 minutes and was recorded and transcribed verbatim. Questions were developed based on the literature reviewed concerning the motivation and emotions of employees. The following areas have been covered during interviews: attitudes towards tipping, impact of tips on emotion and motivation, tipping culture and replication of tips when they are not received. Table 9.1 shows the areas under investigation and the questions asked.

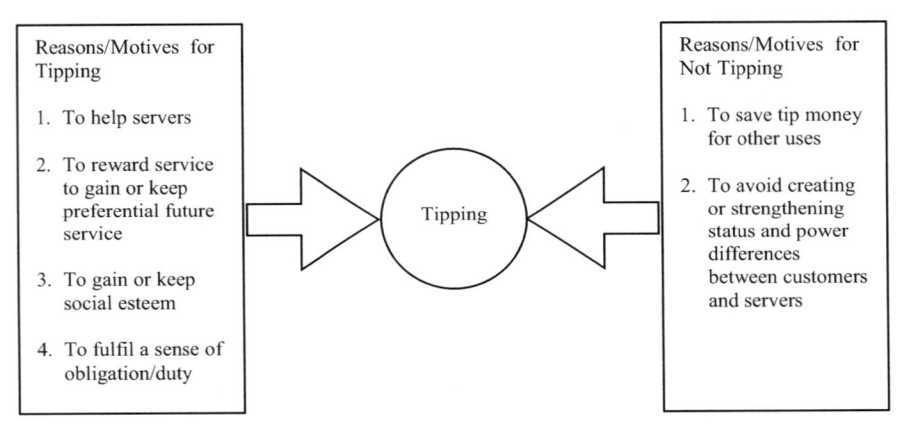

FIGURE 9.1 Tip motives framework.

Source: Lynn, 2015.

TABLE 9.1 Topics investigated and questions asked in the interviews

Topics investigated	Questions being asked
Attitudes towards tipping	What is your attitude to tipping?
	What kind of service did you work in?
	Would you expect to receive tips?
Impact of this on emotion and motivation	If tips were received how would this make you feel and how would it affect the work, you do?
	If tips weren't received how would this make you feel and how would it affect the work, you do?
	Would you ask for a tip?
Tipping culture	Do you think tipping is a good or bad culture in hospitality and what is your reason?
Replication of tip when it is not received	What incentive would you like to see if tips weren't received?

Template Analysis (TeA) approach by King (2004a, b) has been employed for analysing the data. TeA is a technique that analyses data to identify tight generic themes originating within the data, whilst simultaneously allowing for problem-solving and theory-building (Tabari et al., 2020:198). Studies that use TeA for analysing their data are mostly from interview transcripts, reflecting the ubiquity of this form of data collection (e.g., Goldschmidt et al., 2006; Lockett et al. 2012; Slade et al. 2009; Thompson et al. 2010). Furthermore, TeA allows for the inclusion of priori themes identified through engaging with the existing literature and theory within the field (Tabari et al., 2020, Tabari and Chen, 2022; Prokop et al., 2023).

There were five steps in the final analysis of the data so that the final template could be formed: transcribing, familiarisation within the data, input into specialist software (if you are using software), initial template from priori themes and developing the final template (King et al., 2017; 2018). All transcripts were read, and all audio recordings were listened to by both researchers to become familiar with the data. Transcripts were then transferred into the MAXQDA (qualitative analysis software). After all coding had been completed individually by both researchers, they met to agree on the final template and make changes, where necessary. The codes have been labelled, as identified from a certain section of the text which can be descriptive (O'Gorman & MacIntosh, 2015).

9.7 Findings and discussion

For the purpose of this chapter and for providing insight full information and with regard to Tabari et al. (2020), the researchers chose to provide a separate template for each group (experience of working and desire to work in a tipping culture) to avoid assuming that tipping has the same impression in both groups. Final templates are presented in Tables 9.2 and 9.3.

Therefore, to identify the differences and similarities of perceptions among these two groups, a comparison analysis took place between the two templates

TABLE 9.2 Final template, participants with experience working in a tipping culture

Experience of working in a tipping culture	
(Positive)	*(Negative)*
1. Reasons for Tipping	5. Reasons for Tipping
1.1 Non Verbal Appreciation	5.1. Shouldn't be a given
1.2 Rewards for Hard Work	5.2. 5Pressure
1.3 Commonplace	5.3. Unknown wage
1.4 Relatable to Server	5.4. Confusion on receiving end
1.5 Satisfaction	
1.6 Higher Motivator	

(Continued)

TABLE 9.2 (Continued)

Experience of working in a tipping culture

(Positive)	(Negative)

2. Replacement of a Tip
   2.1. Management Appraisal
   2.2. Good Wage
3. Not Receiving Tips
   3.1. Unfair
   3.2. Expression of Negativity
   3.3. Annoyance
   3.4. Demotivated
      3.4.1. Lack of Reward
   3.5. No Affect as not used to
      receiving Tips
4. Tipping as a culture
   4.1. Huge Incentive
   4.2. Improves Employee Morale
   4.3. Motivation and Encouragement
   4.4. Good form of Communication
	6. Tipping as a culture
	   6.1. Employee Loses sight of actual
	      job
	   6.2. Can cause disagreement with
	      colleagues
	   6.3. Can cause confusion on
	      customer's side

TABLE 9.3 Final template, participants who desire to work in a tipping culture

Desire to work in a tipping culture

(Positive)	(Negative)

1. Reasons for tipping
   1.1. Appreciation
   1.2. Commonplace
   1.3. Keep Positive Morale
   1.4. Set Tipping Percentage
2. Replacement of a Tip
   2.1. Other Financial Incentives
      2.1.1. Bonuses
      2.1.2. Good Wage
   2.2. Reasonable Working Hours
   2.3. Good Team Work
   2.4. Good Management Support
   2.5. Non-Monetary Feedback
   2.6. Job Security
	5. Reasons for Tipping
	   5.1. Sympathy
	6. Tipping as Culture
	   6.1. Optional
	      6.1.1. Can create
	         Disappointment
	      6.1.2. Demotivation
	   6.2. Dangerous Technique for
	      Motivation
	   6.3. Reliant on Tips
	      6.3.1. Delivering Service only
	         for Tip

(Continued)

TABLE 9.3 (Continued)

Desire to work in a tipping culture

(Positive)	*(Negative)*

3. Effects on not Receiving Tips
    3.1. Demotivation
    3.2. Expectations of a High Wage
    3.3. Decreases on the Quality of Work
    3.4. Seeks other Form of Motivation
    3.5. Wrong Assumption
        3.5.1. Service was not Satisfactory
    3.6. Assurance
    3.7. Changing Job
4. Tipping as a culture
    4.1. Would Prefer Higher Wage Rather Than Tip
    4.2. As Long as the Individual Employee Received the Tip not the Company
    4.3. Motivation and Financial Incentive
    4.4. Equal
        4.1.1. To Everyone
        4.1.2. Everywhere

to draw a comprehensive conceptual framework from the key findings. As noted by Tabari et al. (2020), direct quotes from interviewees have been used to help the interpretation and provide a more in-depth understanding of data. The coding was accomplished through all sections of the interview text that were possibly relevant to the research question. The 'bottom-up' approach was chosen for coding, with some of the priori themes being developed from the literature. Forty per cent of participants are male, 60% female and have all graduated with a hospitality management degree and working in the service sector. An overview of the participants' demographic information is shown in Figure 9.2.

All participants (with working experience and desire to work) expected to receive tips since they are working in a tipping culture. Participants who had worked in a tipping culture highlighted that 60% of them had not received tips in their most present role in a tipping culture. In this respect, 90% of participants had expected to receive tips, apart from the 10% who stated they had been made aware that this was not the case in their organisation. Van Horn and Schaffner (2003), in their research, stated that waiters and waitresses employed in public bars and restaurants are more likely to receive tips than those in

Participants Numbers	Age	Gender	Nationality
P. 1	23	Female	British
P. 2	22	Male	Irish
p. 3	27	Male	Serbian
p. 4	25	Female	Serbian
p. 5	21	Male	Welsh
P. 6	23	Male	British
p. 7	21	Female	British
P. 8	20	Female	Irish
P. 9	22	Male	British
P. 10	28	Female	Bulgarian
P. 11	21	Female	Cypriot
P. 12	21	Female	British
P. 13	20	Female	British
P. 14	22	Male	British
P. 15	21	Female	British
P. 16	21	Female	British
P. 17	21	Male	British
P. 18	23	Female	Spanish
P. 19	21	Female	British
P. 20	28	Male	Irish

(P. 1–P. 10) Experience of working in a tipping culture

(P. 11–P. 20) Desire to work in a tipping culture

FIGURE 9.2 Participants' demographic information.

private restaurants or bars. In addition to that, the 60% who had not received tips mostly were working in a private club/ organisation and the remaining (40%) who had received tips were working in restaurants or luggage handling in hotels. Parrett (2011) stated that an employee is more likely to receive a tip if they have previous experience of serving to individuals.

9.7.1 Impact of not receiving tips

Participant stated that they would not have been affected by not receiving a tip; however, when the wage is lower than minimum wage, they would feel some other emotional impact and dissatisfaction feeling from their job. However, the participants who desired to work in a tipping culture claimed that they would not be affected if they did not receive a tip; this may be because they have never been in a situation where others may receive a tip and they would not, in an environment that tipping culture is a norm.

One of the participants, who had worked in a tipping culture, stated,

Prior to getting the job I was informed there would be a non-tipping policy, but once I was there working in that culture it seemed only fair to receive tips, it eventually made me feel negative about the work and lead me to feel demotivated at work.

In comparison to this, a few participants who desired to work in a tipping culture replied to the questions 'Would you expect to receive tips?' with 'Yes, it is America and that is known for tipping'.

Employees from non-tipping cultures like the United Kingdom or Serbia may react differently when they are in an environment with different cultures that they are not familiar with. Most of the participants who had worked in a tipping culture said a negative expression was exhibited when a tip was not received. A few participants (30%) stated that 'at first they didn't feel any negativity since they were used to not receive any tips until they had acclimatised to the working culture and gradually, they felt the negativity when they were not receiving tips'. In comparison with participants who desire to work in a tipping culture, 100% stressed that they wouldn't be affected at all by not receiving a tip, and 20% highlighted that if they are the only ones not receiving tips and other colleagues of them receive tips, they would possibly feel bad and less motivated. The negative feeling leads to demotivation at work and may lead to them being unhappy at work. Participants stated that 'I felt unfair when I didn't receive tips and others received tips, I felt unhappy with myself and lost my confidence', 'Gradually I got demotivated and started looking for another job', 'I felt disappointed'. Fairness and equality play an important role as motivators for employees and organisational settings (Folger and Cropanzano, 1998).

9.7.2 Impact of receiving tips

Another key finding from the research was linking tipping to a sense of financial incentive/motivation. Shamir (1983) identified that the tip is an immediate monetary reward which creates a high sense of motivation. The tipping option in service may be one of the factors to motivate employees to provide a higher level of service (Kwortnik et al., 2009). However, participants have highlighted the fact that being motivated by organisations is very helpful in providing a satisfactory feeling from a job for employees, and if tips were not received, then some sort of other motivation would be preferred. One of the participants identified that 'I will consider not accepting the job if no other incentive was offered'. In addition to this, participants highlight the fact that if an organisation does not offer a living wage standard to their employees, not receiving a tip may lead to high staff turnover, since employees will rely on tips more to be able to make ends meet. Receiving a tip for participants of this research who came from a non-tipping culture plays an important role, is great motivation at first and works as a reward element for them, but in some cases, after a while, they feel the conflict and unfairness among employees. Some of them who had a working experience mentioned that 'at first I felt so shock as I was doing my job and got more motivated to work harder', 'I felt being appreciated for my job that I like and motivated me more' and 'I felt appreciated at first but after

TABLE 9.4 Replacement for tipping

Replacement of a tip	Financial	Non-financial
	Bonus	Verbal Appreciation
	Living Wage	Positive Environment
	Overtime Pay	Job Security
		Reasonable Hours

a few months I could feel the conflict among employees and unfairness of the practice in overall'.

It has been highlighted by participants that more non-financial incentives could be seen as a good idea to replace the tip and provide equality to all employees. However, many of the non-financial replacements/ appropriate compensation were only identified after the participants had stated a living wage was necessary to stay in the job; a few examples have been shown in Table 9.4.

9.7.3 Tipping as a good or bad culture

Participants who had worked in a tipping culture agreed that it is a good culture for the hospitality and service industry; however, all participants did state that 'there was a lot of conflict in the workplace over tips among employees'. One respondent identifies that 'some of the staff would get angry when others received tips'. Another respondent states, 'tipping can lead to some employees losing sight of what really matters and what is true service'. Participants believe that 'it led employees to work harder makes them more tired only to get higher tips and may increase negative competition within employees'. When looking at the responses, 100% of participants who had worked in a tipping culture were torn between whether or not they thought it was a positive or negative culture. However, they believed that if the employer provided a 'sharing tips' rule, it would provide an equal opportunity for all employees, those who worked in front and the kitchen, and they could work more as a team, feel appreciation and less conflict among employees. According to the literature, tipping practices within organisations will lead to employee competition in the workplace (Dublanica 2010). Considering that tipping is a voluntary practice and not a mandatory practice it can create a sense of negativity, competition and conflict and affect teamwork in the workplace.

The findings of this study highlighted that the emotions (appreciation, negativity, inequality, equality, satisfaction, dissatisfaction and disappointment), attitudes (motivated, demotivated, conflict, incentive), actions (working harder, looking for another job, conflict, unhealthy competition) and expectations (fairness, equality, same rule) can vary in a tipping culture, and for those who are from a non-tipping culture, these elements will change once experience has been gained working in a tipping culture. In summary, if employees come from a non-tipping culture and start working in a tipping culture at first, they wouldn't be affected by not receiving a tip, since they are not used to it, but after a while working in the environment and assimilating with the culture, they will react to it like other employees. However, if they receive a living wage, then a tip is not needed and customers should not be expected to pay more than the actual bill, which will promote a standard service delivery and there would not be any difference in delivering service when receiving or not receiving a tip.

As King et al. (2018) stressed, the flexible and open nature of qualitative research made it appropriate to the aim of this research and to the intention of developing a conceptual framework to show the visual comparison between two groups and their expectations and reactions towards tipping; therefore, the following conceptual framework has been proposed by the research based on the themes and key findings of data. Therefore, a conceptual framework has been drawn (see Figure 9.3) from the themes that have been developed from the findings, to show the impact of tipping and the perception of those who have worked in a tipping culture and those who desire to work in such an environment with a non-tipping cultural background. The following conceptual framework aimed to visualise the themes from the codes. It shows the perceptions of employees from a non-tipping culture towards receiving or not receiving tips in a tipping culture between those who have worked and those who desire to work in a tipping environment.

The action, emotions and attitude elements were all identified by the participants, which have been discussed earlier on. This framework can be used to

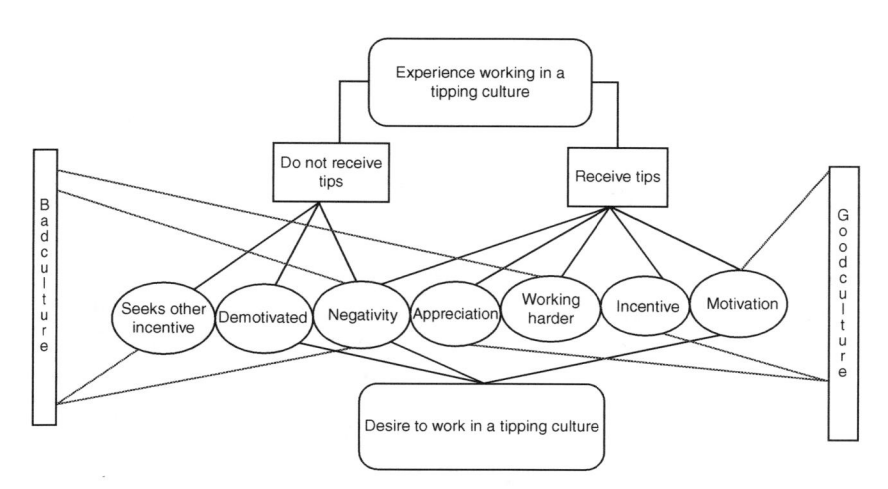

FIGURE 9.3 Conceptual framework, employees' perceptions on tipping from non-tipping culture.

create a more expanded model of the motivational effects on employees and their expectations towards tipping in different cultures and organisations, not only in the hospitality and service industry.

9.8 Practical and managerial implication

The results of this study, according to the exploratory nature of it, do offer some practical and managerial implications. Based on the findings of this study, some useful suggestions have been made for the service sector that could help them overcome the factors that appear to affect employees' expectations and motivations in a workplace with regard to tipping. From a practical point of view, the traditional tipping system and culture among service sectors provide financial help to the employees and motivate them to provide maximum customer service; however, in many cases it leads to conflict and demotivation among employees, which can be solved by practicing a sharing concept, providing a teamwork effort within the group working especially in the restaurant (e.g., waiters and kitchen staff) to avoid unhealthy competition and unnecessary conflict. Furthermore, providing alternative replacements or compensation practices instead of tips will help to improve service standards and employee satisfaction.

Although the findings of this research have highlighted the fact that in some cases tips can be considered a good culture for the service industry, there are still disadvantages of the tipping system that has been seen mostly as a bad culture, that may lead to conflict, lack of teamwork, unhealthy competition and demotivation among employees. This will provide an opportunity for managers, in the service sector, to provide a solution on the tipping system and change the directions towards better practice within organisations. Last, but not least, this study opens up possibilities for developing more research on the impact of tipping on employees and more in-depth research on employees' perceptions of tipping culture and the comparison with a non-tipping culture environment.

9.9 Limitations and future research directions

Aside from the practical and managerial implications, the study has several limitations. The study is exploratory, and no firm predictions were presented. Although the participants of this study have been chosen from multiple non-tipping cultures such as Britain, Ireland, Serbia, Cyprus and Spain, future research could look at other European countries as well as Asian countries such as China and Japan and provide a comparison study between employees who receive tips in a non-tipping cultural environment and the impact of it on employees' expectations and emotions. Further research is needed to explore in what ways can tipping cultures be modified in order to keep employee motivations high. This research has helped to identify that further research is also needed in respect to how to solve the negative sense that tipping brings to a workplace. Ultimately, an analysis of comparisons between tipping motivations between casual or fine dining restaurant customers, full-service restaurants (FSR) and imited-service restaurants (LSR) could help extend the literature even further when looking at the motivations of tipping, including their surroundings as suggested by Whaley et al. (2014). Furthermore, future research can investigate the employees' attitude and provided service towards customers or tourists from a non-tipping culture which often would not give tips.

9.10 Conclusion

In conclusion, this chapter has advanced our understanding of employees from a non-tipping culture and their view of tipping. Most research has examined consumers, reactions and views towards tipping. The result of this study provides new insights towards the impacts of tipping on employees' expectations, motivations and emotions from a non-tipping culture, as well as looking at the effect of receiving or not receiving tips on employees from a non-tipping culture.

References

Abraham, Z. (2014), "The influence of national culture on tipping behaviour". *UNLV Theses, Dissertations, Professional Papers, and Capstones. (Unpublished)*. https://digitalscholarship.unlv.edu/cgi/viewcontent.cgi?article=3627&context=theses dissertations (Accessed on 10 January 2020).

Amer, S. (2002), "Minority report". *Restaurant business*, 27–38.

Azar, O.H. (2004), "Optimal monitoring with external incentives: The case of tipping". *Southern Economic Journal*, 7(1), pp. 170–181.

Azar, O.H. (2005), "Who do we tip and why?, An empirical investigation". *Applied Economics*, 37(16), pp. 1871–1879.

Azar, O.H. (2007), "Why pay extra? Tipping and the importance of social norms and feelings in economic theory". *Journal of Socio-Economics*, 36(2), pp. 250–265.

Azar, O. H. (2011), "Business strategy and the social norm of tipping". *Journal of Economic Psychology*, 32(3), 515–525.

Barkan, R., Erev, I., Zinger, E. and Tzach, M. (2004), "Tip policy, visibility and quality of service in cafes". *Tourism Economics*, 10(4), pp. 449–462.

Bodvarsson, O. and Gibson, W. (1994), "Gratuities and customer appraisal of service: evidence from Minnesota restaurants". *Journal of Socio-Economics*, 23(3), pp. 287–302.

Brewster, Z. W. and Mallinson, C. (2009), "Racial differences in restaurant tipping: A labour process perspective". *The Service Industries Journal*, 29(8), 1053–1075.

Crespi, L. P. (1947). The implications of tipping in America. *Public Opinion Quarterly*, 11(3), pp. 424–435.

Creswell, J. W. and Plano Clark, V.L. (2010), *Designing and conducting mixed methods research* (2nd ed.). Thousand Oaks, CA: Sage.

Cunningham, M.R. (1979), Weather, mood, and helping behavior: Quasi-experiments with the sunshine Samaritan. *Journal of Personality and Social Psychology*, 37, pp. 1947–1956.

Curtis, C.R., Upchurch, R.S. and Severt, D.E. (2009), "Employee motivation and organizational commitment: A comparison of tipped and non tipped restaurant employees". *International Journal of Hospitality & Tourism Administration*, 10(3), pp.253–269.

Dublanica, S. (2010), *Keep the Change: a Clueless Tipper's Quest to Become the Guru of the Gratuity*. Brilliance Publishing.

Ferdman, R. (2016), "I dare you to read this and still feel good about tipping". *The Washington Post*, 18 February. https://www.washingtonpost.com/news/wonk/wp/2016/02/18/i-dare-you-to-read-this-and-still-feel-ok-about-tipping-in-the-united-states/ (accessed 3 June 2020).

Folger, R. and Cropanzano, R. (1998), *Organizational justice and human resource management*. Sage Publications, Inc.

Goldschmidt, D., Schmidt, L., Krasnik, A., Christensen, U. and Groenvold, M. (2006), "Expectations to and evaluation of a palliative home-care team as seen by patients and careers". *Supportive Care in Cancer*, 14(12), pp. 1232–1240.

Gössling, S., Fernandez, S., Martin-Rios, C., Reyes, S.P., Fointiat, V., Isaac, R.K. and Lunde, M. (2020), "Restaurant tipping in Europe: a comparative assessment". *Current Issues in Tourism*, pp. 1–13.

Hemenway, K. (1984), Objects, parts, and categories. *Journal of Experimental Psychology: General*, 113(2), pp. 169–193.

Ivkov, M., Božić, S. and Blešić, I. (2017), "The effect of service staff's verbalized hospitality towards group diner's additional purchases and tipping behaviour". *Scandinavian Journal of Hospitality and Tourism*, 19(1), pp.82–94.

Karabas, I., Orlowski, M. and Lefebvre, S. (2020), What am I tipping you for? Customer response to tipping requests at limited-service restaurants. *International Journal of Contemporary Hospitality Management*, 32(5), pp. 2007–2026.

King, N. (2004a), "Essential guide to qualitative methods in organizational research", in C. Cassell and G. Symon (eds), *Essential Guide to Qualitative Methods in Organizational Research*, London: Sage, pp. 256–270.

King, N. (2004b), "Using templates in the thematic analysis of text". *Essential Guide to Qualitative Methods in Organizational Research*, pp. 256–270.

King, N. and Brooks, J. M. (2017), *Template Analysis for Business and Management Students*, London: Sage, pp. 1–95.

King, N., Brooks, J. M. and Tabari, S. (2018), "Template analysis in business and management research", in M. Ciesieleska and D. Jemielniak (eds), *Qualitative Methodologies in Organization Studies: Volume II: Methods and Possibilities*, London: Palgrave, pp. 179–206.

Koetke, C. (2016), "To Tip or not to tip?" *Fast Casual. News Feature*. http://search.proquest.com.lcproxy.shu.ac.uk/docview/1798843565?accountid=13827&rfr_id=info%3Axri%2Fsid%3Aprimo (Accessed on 27 January 2020).

Koku, P. S. (2005), "Is there a difference in tipping in restaurant versus non-restaurant service encounters, and Abstract do ethnicity and gender matter?" *Journal of Services Marketing*, 19(7), pp. 445–452.

Koku, P. S. and Savas, S. (2016), Restaurant tipping and customers' susceptibility to emotional contagion. *Journal of Services Marketing*, 30(7), pp. 762–772.

Kwortnik, R.J., Jr, Lynn, W.M. and Ross Jr, W.T. (2009), "Buyer monitoring: A means to insure personalized service". *Journal of Marketing Research*, 46(5), pp. 573–583.

Landsburg, S. E. (1993), Rochester Center for.

Lin, I. and Namsivayam, K. (2011), "Understanding restaurant tipping systems: a human resources perspective". *International Journal of Contemporary Hospitality Management*, 23(7), pp. 923–940.

Lockett, S., Hatton, J., Turner, R., Stubbins, C., Hodgekins, J. and Fowler, D. (2012), "Using a semi-structured interview to explore imagery experienced during social anxiety for clients with a diagnosis of psychosis: An exploratory study conducted within an early intervention for psychosis service". *Behavioural and Cognitive Psychotherapy*, 40(1), pp. 55–68.

Lui, C. (2008), "The perceptions of waiters and customers on restaurant tipping". *Journal of Services Marketing*, 19(7), pp 445–452.

Lynn, M. (2000), "National personality and tipping customs". *Personality and Individual Differences*, 28(2), pp. 395–404.

Lynn, M. (2011), "Race differences in tipping: Testing the role of norm familiarity". *Cornell Hospitality Quarterly*, 52(1), pp. 73–80.

Lynn, M. (2015), "Service gratuities and tipping: A motivational framework". *Journal of Economic Psychology*, *46*, pp. 74–88.

Lynn, M. (2016), "Motivations for tipping: how they differ across more and less frequently tipped services". *Journal of Behavioral and Experimental Economics*, 65.

Lynn, M. (2018), "The effects of tipping on consumers' satisfaction with restaurants". *Journal of Consumer Affairs*, 52(3), pp. 746–755.

Lynn, M. and Brewster, Z. W. (2015), "Racial and ethnic differences in tipping: The role of perceived descriptive and injunctive tipping norms". *Cornell Hospitality Quarterly*, 56(1), pp. 68–79.

Lynn, M. and Graves, J. (1996), "Tipping: an incentive/ reward for service?". *Hospitality Research Journal*, 20(1), pp. 1–13.

Lynn, M. and McCall, M. (2000), "Gratitude and gratuity: a meta-analysis of research on the service-tipping relationship". *Journal of Socio-Economics*, 29(2), pp. 203–214.

Lynn, M. and Simons, T. (2000), Predictors of male and female Servers' average tip earnings 1. *Journal of Applied Social Psychology*, 30(2), pp. 241–252.

Lynn, M. and Starbuck, M. M. (2015), "Tipping customs: The effects of national differences in attitudes toward tipping and sensitivities to duty and social pressure". *Journal of Behavioral and Experimental Economics*, 57, pp. 158–166.

Lynn, M. and Wang, S. (2013), "The indirect effects of tipping policies on patronage intentions through perceived expensiveness, fairness and quality". *Journal of Economic Psychology*, 39(1), pp. 62–71.

Lynn, N. and Latane, B. (1984), "The psychology of restaurant tipping". *Journal of Applied Social Psychology*, 14(6), pp. 549–561.

Mankiw, G. (2007), No, really, it's up to you. *Greg Mankiw's Blog*, *1*.

Margalioth, Y., Sapriti, A. and Coloma, G. (2010), "The social norm of tipping, its correlation with inequality, and differences in tax treatment across countries". *Theoretical Inquiries in Law*, 11(2), pp. 561–588.

Matilla, A. and Enz, C. (2002), "*The role of emotions in service encounters*", *Journal of Service Research*, 4(4), pp. 268–277.

Mccall, M. and Lynn, A. (2009), "Restaurant servers' perceptions of customer tipping intentions". *International Journal of Hospitality Management*, 28(4), pp. 594–596.

Namasivayam, K. and Upneja, A. (2007), Employee preferences for tipping systems. *Journal of Foodservice Business Research*, 10(2), pp. 93–107.

Neuman, W. L. (2017), *Understanding Research* (1st ed.). Pearson.

Noll, E. D. and Arnold, S. (2004), "Racial differences in restaurant tipping: Evidence from the field". *The Cornell Hotel and Restaurant Administration Quarterly*, 45(1), 23–29.

O'Gorman, K. and MacIntosh, R. (2015), *Research Methods for Business and Management a Guide to Writing Your Dissertation*. 2nd ed. Goodfellow Publishers Ltd.

Parrett, M. (2011), "Do people with food service experience tip better?" *The Journal of Socio-Economics*, 40(5), pp. 464–471.

Prokop, D., Tabari, S. and Chen, W. (2023), Survival instincts of Chinese entrepreneurs in the UK: Adaptation or hibernation. *International Journal of Entrepreneurship and Small Business*.

Rokou, T. (2014), Americans Most Gratuitous When It Comes to Tipping on Vacation, According To TripAdvisor Survey. https://www.tripadvisor.com/PressCenter-i6820-c1-Press_Releases.html (Accessed 24 January 2020).

Saayman, M. and Saayman, A. (2015), "Understanding tipping behaviour – An economic perspective". *Tourism Economics*, 21(2), pp. 247–265.

Saunders, S. G. (2015), "Service employee evaluations of customer tips: An expectations-disconfirmation tip gap approach". *Journal of Service Theory and Practice*, 25(6), pp. 796–812.

Segrave, K. (2009), *Tipping: An American Social History of Gratuities*. McFarland.

Seiter, J.S. and Weger, Jr., H. (2010), "The effect of generalized compliments, sex of server, and size of dining party on tipping behavior in restaurants". *Journal of Applied Social Psychology*, 40(1), pp.1–12.

Shamir, B. (1983), "A note on tipping and employee perceptions and attitudes." *Journal of Occupational Psychology*, 56(3), pp. 255–259.

Shamir, B. (1984), "Between gratitude and gratuity: an analysis of tipping". *Annals of Tourism Research*, 11(1), pp. 59–78.

Slade, P., Haywood, A. and King, H. (2009), "A qualitative investigation of women's experiences of the self and others in relation to their menstrual cycle". *British Journal of Health Psychology*, 14(1), pp. 127–141.

Tabari, S. and Chen, W. (2022), Ethnic female entrepreneurs in the service sector: challenges and motivations. In *Global Strategic Management in the Service Industry: A Perspective of the New Era* (pp. 99–118). Emerald Publishing Limited.

Tabari, S., King, N. and Egan, D. (2020), "Potential application of template analysis in qualitative hospitality management research". *Hospitality & Society*, 10(2), pp. 197–216.

Thompson, J. L., Jago, R., Brockman, R., Cartwright, K., Page, A. S. and Fox, K. R. (2010), "Physically active families – De-bunking the myth? A qualitative study of family participation in physical activity". *Child: Care, Health and Development*, 36(2), pp. 265–274.

US Census Bureau (2017), "2017 NAICS definition 722511 full-service restaurants". www.census.gov/cgi-bin/sssd/naics/naicsrch?code=722511&search=2017%20NAICS%20Search (accessed 20 December 2019).

Van Horn, C. E. and Schaffner, H. A. (2003), *Work in America: An Encyclopedia of History, Policy, and Society*. ABC-CLIO.

Were, S. O., Miricho, N. M. and Maranga N. V. (2019), "A review of hospitality's restaurant operations: An explanation of the tipping phenomenon". *Global Journal Of Management And Business Research*, 19(5), pp.52–55.

Whaley, J. E., Douglas, A. C. and O'Neill, A. O. (2014), "What's in a tip? The creation and refinement of a restaurant-tipping motivations scale: a consumer's perspective". *International Journal of Hospitality Management*, 37, pp. 121–130.

10

LUXURY BUYING BEHAVIOUR ACROSS DIFFERENT TOURISTS' GENERATIONS

David D'Acunto

The evolutionary nature of luxury: From 'owing' to 'being'

While till the end of the last century, the main luxury consumption was concentrated on products pertaining to the fashion, jewellery, perfumery, cosmetics or automotive industries, today a clear trend emerges towards new contexts such as tourism, hospitality and travel (Chevalier & Mazzalovo, 2021; Amatulli & Guido, 2011). The concept of luxury in tourism and hospitality is going through a radical change, with most of the current research started considering it under a multidimensional framework by acknowledging the complexity in the dynamics driving luxury consumers (Iloranta, 2022; Kauppinen-Räisänen et al., 2019). The traditional view of luxury revolving around the concepts of conspicuous consumption and gaining status (Nueno & Quelch, 1998) is now outdated and replaced by a new way of interpreting luxury that rather focuses on how luxury is experienced (Thomsen et al., 2020). What forms value for consumer seeking luxury is now the experience.

Value for consumer is an 'interactive relativistic preference experience' (Holbrook, 1999, p. 5). According to this conceptualisation, what is valuable for consumers (i.e., the value) is what can be generated by the interaction between an individual (i.e., subject) and a product, service or experience (i.e., object), thus revealing value as a relativistic concept, driven by personal, situational, comparative and preferential differences. Consumer value therefore does not directly refer to the brand or the object itself, but rather derives from the experience. With regard to the luxury domain, delivering experiences for a brand means offering consumers a combination of social, emotional and symbolic value (Gupta et al., 2023), in addition to the functional one. Consumers expected outcomes of consuming luxury experience are feeling comfort,

DOI: 10.4324/9781032637778-11

well-being and sense of belonging (Han & Kim, 2020; Klaus & Tarquini-Poli, 2022). This multidimensional understanding is key to study luxury consumer behaviour and to shed light on the different value drivers, with the aim of providing specific guidance to organisations on how to improve their products and services and the related marketing campaigns in the luxury travel domain.

A main distinction between luxury and generalist brands relies in the predominance of non-functional versus functional value in the first ones (Vigneron & Johnson, 1999). Luxury brands deliver a specific value proposition centred on social and emotional uniqueness. Nowadays luxury firms and customers engage different value co-creating processes allowed by the new technologies. By stimulating two-way interactions with their customers, luxury firms can benefit of symbolic, experiential and relational values that are considered key differentiating elements for their luxury brands (Tynan et al., 2010). Avoiding the separation between production and consumption, like it occurs in traditional contexts, luxury experiences can create value for customers.

Consumers' search for value has moved luxury from 'having' to 'being' (Figure 10.1) that is from the concept of possessing to experiencing (Cristini et al., 2017; Cohen et al., 2022). Experiences are more closely related to the self and are not replicable from person to person compared to material goods. As a result, experiences have boosted the global increase in the consumption of luxury services (Carter & Gilovich, 2012). Luxury travel experiences can be easily shared online as a way of tourists' expression with the threefold aim of

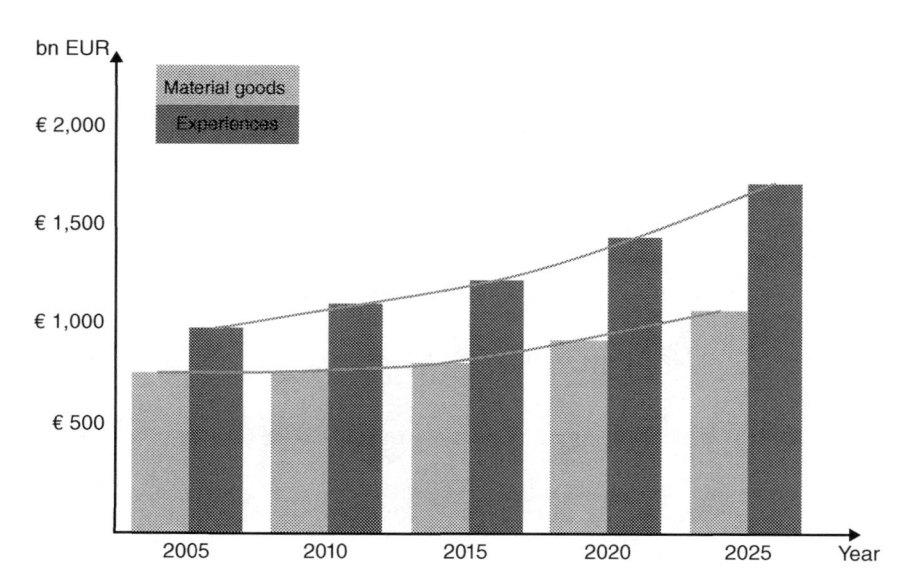

FIGURE 10.1 Total annual expenditure on enrichment vs. material goods.

Source: Own elaboration on Oxford Economics data.

building desirable self-image, enhancing self-esteem and gain external validation (Liu & Li, 2021). Luxury experiences include time, space, authenticity,
community, individuality and well-being experiences (Yeoman & McMahon-
Beattie, 2018) and different activities such as luxury travel, fine dining, exclusive tourist experiences and destinations (Kim, 2018).

Following the rise of the experience economy (Thomsen et al., 2020), extant
literature started acknowledging the key role of the experiential value in driving consumers towards luxury choices. Nevertheless, the evolutionary nature
of luxury is nowadays shifting to a deeper level, with new luxury customers
reshaping the industry for instance moving towards sharing economy options
and seeking emotional connection online through social media. The experience
of choosing travelling luxury is much more than just providing some utility, but
is rather showing or signalling to others offline/online a self-interest or status
(Bronner & de Hoog 2019). Similarly, sharing luxury destinations experiences
online leads peers to desire similar travel experiences (Liu et al., 2019). Such
transformation is driven by a generational shift with Millennials and Generation
Z consumers now driving the growth of global luxury revenues. Such new luxury segments of consumers are more environmentally and socially aware and
conscious, thus having higher expectations towards luxury brands and demanding companies' higher involvement in social and environmental responsibility
and in the transparency of their best practices and communication.

The reasons behind luxury tourism consumer choices are therefore not only
the tangible and concrete aspects such as quality, service, price or authenticity,
but also social and individual attributes (Chang et al., 2016; Tynan et al., 2010).
Extant literature confirms how the consumption of luxury takes on symbolic
traits and express specific social and cultural values (Wiedmann et al., 2009;
Sung et al., 2015; Becker et al., 2018; D'Acunto & Volo, 2021).

The traditional concept of tourism, as consumption of products and services tailored for 'mass' tourists, is thus overcome by a new form of tourism
experience, where individuals are actively contributing in forming their emotions, memory, stories and feelings (Atwal & Williams, 2017). As a result and in
response to the new segments and needs of travellers seeking for experiences, an
increasing offer of thematic services and experiences has been dominating the
tourism industry offer (Heyes et al., 2015; Moscardo & Benckendorff, 2010).

Luxury buying behaviour: From goods and services to experiences

Luxury buying behaviour in tourism has seen a notable shift from goods and
services to an emphasis on luxury experiences. This transformation is a
response to changing consumer preferences, values and aspirations, and it
reflects a broader shift in the luxury industry as a whole.

Traditionally, luxury consumption in tourism and hospitality was characterised by the acquisition of high-end goods and services such as designer

clothing, luxury cars and fine dining at exclusive restaurants (Chevalier & Mazzalovo, 2021). Luxury travellers sought opulent accommodations, gourmet cuisine and access to exclusive amenities and facilities in their pursuit of status and prestige (Vigneron & Johnson, 2004).

However, as consumer values and priorities have evolved, there has been a noticeable shift in luxury buying behaviour from the accumulation of material possessions to the accumulation of experiences. This shift is influenced by several factors, including the desire for unique and memorable moments, the pursuit of personal fulfilment and a growing emphasis on sustainability and ethical consumption (Cohen et al., 2014).

Experiential luxury has become a dominant trend in the luxury tourism and hospitality industry (Yang & Mattila, 2016). Luxury consumers are increasingly seeking travel experiences that provide them with emotional, sensorial and cognitive gratification (Shahid & Paul, 2022). Instead of merely possessing luxurious items, they want to immerse themselves in exclusive experiences that leave lasting memories. Luxury hotels, for example, have adapted to this trend by offering personalised and immersive experiences that go beyond the traditional notions of luxury (Bharwani & Mathews, 2021). Guests can now participate in activities such as private wine tastings with renowned sommeliers, exclusive cultural tours or spa treatments tailored to their individual preferences (Anderson, & Harris, 2019). The digital age has played a pivotal role in the shift towards experiential luxury consumption (von Wallpach et al., 2020). Technology and social media platforms have enabled travellers to share their luxury experiences with a global audience, thereby amplifying the value of the experience itself (Cohen et al., 2022). Luxury consumers, particularly younger generations like Millennials and Generation Z, are motivated to curate and share their travel experiences on platforms such as Instagram and TripAdvisor (Park & Lee, 2024). This phenomenon has spurred luxury providers to create 'Instagrammable' moments and shareable experiences that resonate with the desire for social recognition and validation (Michaelidou et al., 2022). Luxury hotels, in particular, invest in creating visually stunning interiors, breathtaking views, and unique design elements to cater to this trend (Jin & Phua, 2014).

Another significant driver of the shift towards experiential luxury is the growing emphasis on sustainability and ethical consumption (Volo & D'Acunto, 2021). Luxury travellers are increasingly concerned about the environmental and social impacts of their choices, leading them to seek out sustainable and responsible luxury experiences (Han et al., 2018). Luxury hotels and resorts are responding by implementing eco-friendly practices, from energy-efficient building designs to sourcing local, organic and sustainable ingredients for their restaurants (Peng & Chen, 2019). Sustainability certifications and transparent communication of sustainability efforts have become essential components of marketing strategies for luxury accommodations (Berezan et al., 2016).

The shift from goods and services to experiential luxury has profound implications for both luxury providers and consumers. When individuals are travelling they show different preferred timing of consumption for experiential and material purchases (Hwang et al., 2019). Luxury businesses must continuously innovate to create unique and memorable experiences that cater to the evolving desires of luxury travellers (Atwal & Williams, 2017). Such shift is a response to changing consumer values and aspirations. Over the last decades, research illustrated how luxury tourism and hospitality have embraced this shift towards experiential luxury, offering travellers the opportunity to create lasting memories and seek personal fulfilment through immersive and sustainable experiences. This transformation not only challenges luxury providers to innovate continually but also highlights the importance of understanding and adapting to evolving consumer preferences in the dynamic world of luxury travel and hospitality.

Luxury across generational cohorts (comparing Gen Z, Gen Y, Gen X and Baby Boomers)

Luxury has evolved and been redefined across various generations (Smith & Luxmore, 2017). As the world has seen shifts in socio-economic trends, technological advancements and cultural movements, the perception and consumption of luxury have been moulded accordingly. This section provides a general overview across different generational cohorts and discusses their peculiarities in approaching luxury within the tourism and hospitality domain.

Gen Z

Gen Z, born between 1997 and 2012, represents the youngest generation of tourists, and their approach to luxury consumption is distinct. They have grown up in a digital age, which has profoundly shaped their preferences and expectations. For Gen Z travellers, luxury is often defined by experiences over possessions. They seek unique and Instagram-worthy experiences that can be shared on social media platforms. Luxury, for them, is about creating memories and curating their online personas.

This generation is also highly conscious of sustainability and ethical considerations. They are more likely to choose eco-friendly and socially responsible luxury options, such as eco-resorts or hotels with strong sustainability practices. Brands that align with Gen Z's values and offer immersive, tech-savvy, and sustainable experiences are likely to capture their attention.

Gen Z travellers are also characterised by a desire for personalisation and convenience. They expect seamless digital booking processes, mobile check-ins and tailored recommendations. Luxury brands that can offer personalised services and tech-driven experiences are more likely to resonate with Gen Z tourists.

This digital-native generation, having grown up in an era dominated by social media and rapid technological advancements, perceives luxury in a much more experiential and purpose-driven manner (Parker, 2020). For them, luxury is not just about owning an expensive item but about the experience and story behind it. Sustainable luxury brands and those that promote inclusivity resonate more with this generation. They are looking for authenticity, and brands that can offer a genuine connection or stand for a cause are seen as luxurious (Johnson & Turner, 2019).

Gen Y or Millennials

Millennials, born between 1981 and 1996, have had a significant influence on the luxury travel industry. They prioritise experiences and are willing to spend on travel and hospitality that promise unique and authentic encounters. Millennials are known for their desire to explore off-the-beaten-path destinations, engage in cultural immersion and embrace adventure.

Luxury for Millennials often means boutique hotels, experiential dining and activities that allow them to connect with local culture. They value authenticity and seek out destinations and accommodations that provide a genuine sense of place. Brands that offer cultural experiences and opportunities for meaningful connections are attractive to this generation.

Millennials are also the first generation to fully embrace the sharing economy, with platforms like Airbnb becoming a popular choice for accommodations. Luxury brands in the travel sector have adapted by offering home-sharing options and experiences that cater to this trend. Millennials, having experienced both the pre-digital and digital era, have a unique perspective on luxury. While they appreciate the traditional symbols of luxury, they also value experiences over possessions (Harris & Gilbert, 2018). Travel, for instance, is seen as a luxury. Brands that can offer personalised experiences or have a strong online presence appeal to this cohort. Furthermore, their purchasing decisions are also influenced by peer reviews and online influencers (Brown & Hayes, 2016).

Gen X

Gen X, born between 1965 and 1980, represents a generation with a unique blend of values and preferences when it comes to luxury travel. They often prioritise comfort, quality and relaxation in their travel choices. Gen X travellers are more likely to opt for luxury resorts, cruise vacations and destinations known for their leisurely pace.

For Gen X, luxury often means indulgence and escape from the demands of everyday life. They value high-end amenities, spa services and fine dining. Brands that can provide a sense of indulgence and cater to their desire for relaxation are well-positioned to attract Gen X tourists. This generation also values family travel, making multigenerational trips a common occurrence.

Luxury brands that can accommodate the needs and preferences of different family members while still offering a sense of exclusivity are likely to be favoured by Gen X travellers. Growing up in an era of economic fluctuations, this generation has a pragmatic approach to luxury. While they do appreciate the finer things in life, they also look for value and longevity in luxury products (Stevens, 2017). For Gen X, luxury is also about exclusivity and a sense of belonging to an elite group. They are more likely to be brand loyal compared to the younger generations and place importance on the heritage and legacy of luxury brands (Miller & Rose, 2015).

Baby Boomers

Baby Boomers, born between 1946 and 1964, represent an older yet influential segment of luxury travellers. They often have more disposable income and free time for travel. Luxury for Baby Boomers often means traditional luxury experiences, such as staying in renowned hotels, taking luxury cruises and indulging in fine dining. This generation places a high value on service, reliability and reputation. Luxury brands that have a long-standing and impeccable track record are appealing to Baby Boomers. They seek out destinations and accommodations that offer comfort, familiarity and a sense of nostalgia.

Moreover, Baby Boomers are more likely to prioritise the cultural and historical aspects of travel. They enjoy exploring heritage sites, museums and attending cultural events. Luxury brands that offer cultural enrichment and educational experiences can resonate with this generation.

Baby Boomers, having experienced post-war prosperity, have a more traditional view of luxury. For them, luxury products are often seen as status symbols and represent success (Davis & Ward, 2013). They are more inclined towards established luxury brands and are less influenced by digital trends. However, they also value craftsmanship, quality and the history behind luxury items (Smith & Luxmore, 2017).

Luxury is therefore a multifaceted concept that has been perceived differently across generations. While Baby Boomers and Gen X might lean towards the traditional symbols of luxury, Millennials and Gen Z are reshaping the luxury market with their emphasis on experiences, authenticity and purpose-driven brands. As brands navigate the evolving landscape of luxury, understanding these generational nuances is crucial to stay relevant and appealing.

The role of sustainability and ethical dimensions of luxury consumption across generational cohorts

Each generation approaches sustainability and ethical concerns differently, impacting their choices in luxury travel and hospitality. Gen Z is often referred as the 'green' generation, as they prioritise sustainability and ethical practices in their consumption decisions. They are deeply concerned about environmental and social issues and seek out luxury brands and experiences that align with their values. Gen Z tourists are more likely to choose eco-friendly accommodations, engage in responsible tourism practices and support brands that demonstrate commitment to sustainability. In the modern age, sustainability and ethical considerations have emerged as significant influencers in the realm of luxury consumption. With increasing global awareness of environmental, social and economic challenges, consumers are becoming more discerning about their luxury purchases. The role of sustainability and ethics in luxury consumption, however, varies across different generational cohorts: Gen Z, Gen Y (Millennials), Gen X and Baby Boomers. This exploration seeks to understand these variations and the implications for luxury brands.

Traditionally, luxury was synonymous with opulence, rarity and often, excess. However, the 21st century has seen a shift, where luxury is increasingly associated with sustainability and ethical production (Greenwood & Schmidt, 2019). From ethically sourced diamonds to eco-friendly fashion, luxury brands are realising that sustainable practices are not just ethically right but are also becoming a consumer demand. The following section provides a discussion of the perception of sustainability and related practices across generations in the luxury tourism domain.

Gen Z

Arguably the most environmentally conscious generation, Gen Z has grown up amidst discussions of climate change, ethical consumerism and sustainability (Sakdiyakorn et al., 2021). For them, the appeal of a luxury product is significantly enhanced if it aligns with sustainable practices. They are well-informed and often research a brand's ethical stance before making a purchase. Luxury, for Gen Z, is not just about the product but the story behind it – how it's made, who made it and its environmental footprint.

Luxury brands that adopt eco-friendly initiatives, such as reducing carbon emissions, minimising single-use plastics and supporting local communities, are more likely to resonate with Gen Z travellers. These tourists actively seek out experiences that allow them to make a positive impact on the destinations they visit.

Gen Y (Millennials)

Millennials, much like Gen Z, place a high value on sustainability. However, their approach is more balanced between traditional luxury values and modern ethical considerations (Jones & Thompson, 2021). They appreciate the craftsmanship and heritage of traditional luxury brands but expect these brands to evolve in line with contemporary ethical standards. For instance, a Millennial might invest in a high-end fashion brand but would expect the brand to have transparent supply chains and sustainable materials.

They are more inclined to support businesses that prioritise responsible and sustainable practices. Millennials often seek out eco-luxury accommodations, organic and locally sourced dining options, and experiences that have a minimal environmental footprint.

The desire for authenticity extends to their ethical considerations. Millennials are more likely to engage in voluntourism, where they can participate in community-based projects and contribute positively to the destinations they visit. They are also conscious of their carbon footprint and may opt for eco-conscious modes of transportation.

Luxury brands that integrate sustainability into their offerings and actively communicate their ethical initiatives are attractive to Millennials. Transparent reporting on their sustainability efforts and partnerships with organisations focused on social and environmental causes resonate with this generation.

Gen X

This generation, having witnessed the dawn of environmental movements and the evolution of luxury, often finds itself torn between the allure of traditional luxury and the emerging ethical standards (Martin & Lewis, 2022). While they do value sustainability, their purchase decisions in the luxury segment are a blend of brand legacy, perceived value and ethical considerations. They might prioritise sustainability when it aligns with other factors like quality and brand reputation, convenience and comfort. They are more likely to choose luxury options that offer responsible practices but may not actively seek out eco-friendly accommodations or activities. Gen X tourists appreciate luxury brands that incorporate sustainability without compromising on quality or comfort. They value efforts such as energy-efficient facilities, waste reduction and conservation programmes. Luxury brands can appeal to Gen X by seamlessly integrating sustainable practices into the overall guest experience.

Baby Boomers

Growing up in a post-war era, Baby Boomers have a more traditional view of luxury. For them, luxury items are symbols of success and achievement (Davis & Peterson, 2018). While they are becoming increasingly aware of sustainability issues, their luxury consumption patterns are less influenced by these factors compared to younger generations. That said, as global sustainability issues become more pronounced, a segment of Baby Boomers is beginning to align their luxury purchases with brands that showcase ethical responsibility.

Baby Boomers, like Gen X, appreciate sustainability and ethical practices but may not make them the primary criteria for their luxury travel choices. They are more likely to choose options that offer responsible practices when presented to them but may not actively seek out eco-luxury options. Luxury brands that prioritise sustainability and ethical considerations can attract Baby

Boomers by highlighting their initiatives in marketing materials and on their websites. Offering responsible choices as part of package options can also resonate with this generation.

Future challenges for luxury travel and tourism industry across generations

The varying perspectives across generations present both challenges and opportunities for luxury brands. Brands need to strike a balance between maintaining their legacy and adapting to the ethical demands of modern consumers (Smith & Andrews, 2019). Transparency, responsible sourcing and sustainable practices are no longer optional; they are crucial for brand reputation and consumer trust. Moreover, with the purchasing power gradually shifting towards the younger generations, brands need to integrate sustainability into their core values and communication strategies. The role of sustainability and ethical dimensions in luxury consumption is undeniable and varies across generational cohorts. The modern luxury market landscape requires brands to be agile, authentic and aligned with global ethical values. As the definition of luxury evolves, incorporating sustainability will not only be an ethical imperative but also a business one.

Understanding the nuances of luxury buying behaviour across different generations of tourists is a complex but essential task for marketers and businesses operating in the travel and hospitality industry. Each generation brings unique attitudes, values, preferences and behaviours when it comes to luxury consumption while traveling.

Luxury buying behaviour across different generations of tourists reflects a rich tapestry of preferences and values. Gen Z seeks tech-savvy, sustainable and personalised experiences, while Millennials prioritise authenticity and cultural immersion. Gen X values comfort and relaxation, and Baby Boomers appreciate traditional luxury and cultural enrichment. To succeed in the luxury travel and hospitality industry, businesses must tailor their offerings to cater to the unique desires and expectations of each generation, ensuring that they provide experiences that align with their values and preferences. In today's digital age, technology and social media play a pivotal role in shaping the luxury buying behaviour of tourists across different generations. Each generation interacts with technology and social platforms in distinct ways, influencing how they discover, choose and share their luxury travel experiences.

Among the others, it is worth mentioning the following dimensions that will characterise future challenges for luxury tourism providers in approaching different generations of customers.

Digital influence and personalisation

Gen Z and Millennials are often termed as 'digital natives' and 'digital pioneers', respectively. The online realm, particularly social media, has been instrumental in shaping their perceptions of luxury. Luxury for them is no longer confined to physical items. A personalised online shopping experience or an Instagram-worthy travel destination is considered just as luxurious, if not more, than a high-end handbag or watch. They value brands that can offer bespoke experiences, tailored to their individual preferences (Williams & Green, 2022).

Craftsmanship and heritage

While the younger generations are leaning towards the digital experience of luxury, Gen X and Baby Boomers have a deep-rooted appreciation for craftsmanship and heritage. They are more inclined to invest in a luxury item that boasts of its centuries-old crafting techniques or has a legacy behind its name. They seek brands that have a story to tell, a history that adds value to the product in hand (Kapferer & Michaut, 2019).

The shift towards sustainable luxury

One of the most significant shifts in the luxury industry has been the emphasis on sustainability. This trend is particularly driven by Gen Z and Millennials who are more environmentally conscious. They are willing to invest in luxury brands that are ethically produced, have a minimal environmental footprint and contribute positively to society. Luxury is no longer just about exclusivity; it is about responsible consumption (Anderson, & Brown, 2019).

Exclusivity vs. inclusivity

The traditional notion of luxury revolved around exclusivity. However, with the rise of Gen Z and Millennials, there is a push towards more inclusivity. They seek brands that are diverse, promote equality and resonate with global values. While they appreciate the exclusivity that luxury brands offer, they want these brands to be more accessible and in tune with global issues (Peters & Lang, 2021).

Conclusion

As the perception of luxury evolves, brands need to be agile and adaptive. They need to strike a balance between maintaining their legacy and adapting to the digital age. The evolving landscape of luxury consumption, set against a backdrop of rapid technological advancements, global socio-economic shifts and growing environmental consciousness, has revealed distinct generational nuances in perceptions and expectations.

At its core, luxury remains a symbol of exclusivity, opulence and a reflection of personal and societal values. However, as revealed in our exploration, the dimensions that define these values are changing. Luxury is no longer just about possessing an item of rarity or value; it's about the experiences, the stories, the ethical considerations and the personal connections these items or experiences engender.

As luxury brands and businesses strive to remain competitive and relevant, they must tailor their marketing strategies, brand management and sustainability efforts to cater to the specific needs and desires of these diverse generational segments, thus ensuring continued success in the luxury tourism and hospitality market (Smith & Garcia, 2021). The future of experiential luxury in tourism and hospitality lies in customisation, personalisation and the integration of technology to enhance the guest experience further (Buehring & O'Mahony, 2019; Shahid & Paul, 2022). Augmented reality (AR) and virtual reality (VR) technologies, for example, hold immense potential for creating immersive and interactive experiences (Loureiro, 2020). Research should explore how these technologies can be harnessed to enhance the luxury experience, from virtual property tours to immersive destination previews.

References

Amatulli, C., & Guido, G. (2011). Determinants of purchasing intention for fashion luxury goods in the Italian market: A laddering approach. *Journal of Fashion Marketing and Management: An International Journal, 15*(1), 123–136.

Anderson, P. & Brown, L. (2019). Personalization and customization: Trends in luxury travel. *Journal of Luxury Travel Insights, 27*(3), 101–115.

Anderson, P., & Harris, J. (2019). Sustainable luxury in travel: Meeting the ethical demands of the modern traveler. *Eco-Travel Journal, 6*(1), 28–39.

Atwal, G., & Williams, A. (2017). Luxury brand marketing–the experience is everything!. *Advances in Luxury Brand Management*, 43–57. https://doi.org/10.1007/978-3-319-51127-6_3

Becker, K., Lee, J. W., & Nobre, H. M. (2018). The concept of luxury brands and the relationship between consumer and luxury brands. *Kip Becker, Jung Wan Lee, Helena M. Nobre/Journal of Asian Finance, Economics and Business, 5*(3), 51–63.

Berezan, O., Yoo, M., & Christodoulidou, N. (2016). The impact of communication channels on communication style and information quality for hotel loyalty programs. *Journal of Hospitality and Tourism Technology, 7*(1), 100–116.

Bharwani, S., & Mathews, D. (2021). Techno-business strategies for enhancing guest experience in luxury hotels: a managerial perspective. *Worldwide Hospitality and Tourism Themes, 13*(2), 168–185.

Bronner, F., & De Hoog, R. (2019). Comparing conspicuous consumption across different experiential products: Culture and leisure. *International Journal of Market Research, 61*(4), 430–446.

Brown, A., & Hayes, N. (2016). The influence of social media on young consumers' purchasing decisions. *Journal of Marketing Trends, 3*(2), 19–29.

Buehring, J., & O'Mahony, B. (2019). Designing memorable guest experiences: Development of constructs and value generating factors in luxury hotels. *Journal of Hospitality and Tourism Insights, 2*(4), 358–376.

Carter, T. J., & Gilovich, T. (2012). I am what I do, not what I have: the differential centrality of experiential and material purchases to the self. *Journal of personality and social psychology, 102*(6), 1304.

Chevalier, M., Mazzalovo, G., & Boselli, M. (2008). *Luxury Brand Manage- Ment in Digital and Sustainable Times*, 4th ed., John Wiley & Sons, West Sussex.

Cohen, S., Liu, H., Hanna, P., Hopkins, D., Higham, J., & Gössling, S. (2022). The rich kids of Instagram: Luxury travel, transport modes, and desire. *Journal of Travel Research, 61*(7), 1479–1494.

Cohen, S. A., Prayag, G., & Moital, M. (2014). Consumer behaviour in tourism: Concepts, influences and opportunities. *Current issues in Tourism, 17*(10), 872–909.

Cristini, H., Kauppinen-Räisänen, H., Barthod-Prothade, M., & Woodside, A. (2017). Toward a general theory of luxury: Advancing from workbench definitions and theoretical transformations. *Journal of Business Research, 70*, 101–107.

D'Acunto, D., & Volo, S. (2021). Cultural traits in the consumption of luxury hotel services: An exploratory analysis through online reviews data. In *Information and Communication Technologies in Tourism 2021: Proceedings of the ENTER 2021 eTourism Conference*, January 19–22, 2021 (pp. 269–279). Springer International Publishing.

Davis, R. L., & Peterson, M. (2018). Baby Boomers and luxury: Traditional values in a modern world. *Boomer Market Trends, 9*(2), 15–25.

Davis, R. L., & Ward, P. (2013). The era of conspicuous consumption: Understanding the Baby Boomers' luxury consumption patterns. *Journal of Business Research, 66*(7), 1032–1039.

Greenwood, M., & Schmidt, L. (2019). The evolution of luxury: From opulence to sustainability. *Journal of Luxury Studies, 12*(1), 7–19.

Gupta, D. G., Shin, H., & Jain, V. (2023). Luxury experience and consumer behavior: a literature review. *Marketing Intelligence & Planning, 41*(2), 199–213.

Han, S. L., & Kim, K. (2020). Role of consumption values in the luxury brand experience: Moderating effects of category and the generation gap. *Journal of Retailing and Consumer Services, 57*, 102249.

Han, W., McCabe, S., Wang, Y., & Chong, A. Y. L. (2018). Evaluating user-generated content in social media: an effective approach to encourage greater pro-environmental behavior in tourism?. *Journal of Sustainable Tourism, 26*(4), 600–614.

Harris, J., & Gilbert, D. (2018). Luxury in the age of Millennials: Experiences over materialism. *Journal of Luxury Research, 2*(1), 25–35.

Heyes, A., Beard, C., & Gehrels, S. (2015). Can a luxury hotel compete without a spa facility?–Opinions from senior managers of London's luxury hotels. *Research in Hospitality Management, 5*(1), 93–97.

Holbrook, M. B. (Ed.). (1999). *Consumer value: A framework for analysis and research.* Psychology Press.

Hwang, E., Kim, J., Lee, J. C., & Kim, S. (2019). To do or to have, now or later, in travel: Consumption order preference of material and experiential travel activities. *Journal of Travel Research, 58*(6), 961–976.

Iloranta, R. (2022). Luxury tourism–a review of the literature. *European Journal of Tourism Research, 30*, 3007–3007.

Jin, S. A. A., & Phua, J. (2014). Following celebrities' tweets about brands: The impact of twitter-based electronic word-of-mouth on consumers' source credibility perception, buying intention, and social identification with celebrities. *Journal of advertising, 43*(2), 181–195.

Johnson, L., & Turner, R. (2019). Gen Z and the rise of purpose-driven luxury. *Trends in Luxury Marketing*, *4*(1), 12–20.

Jones, R., & Thompson, L. (2021). Millennials: Balancing heritage and ethics in luxury consumption. *Millennial Market Review*, *10*(3), 45–56.

Kapferer, J. N., & Michaut, A. (2019). Are millennials really redefining luxury? A cross-generational analysis of perceptions of luxury from six countries. *Journal of Brand Strategy*, *8*(3), 250–264.

Kauppinen-Räisänen, H., Gummerus, J., von Koskull, C., & Cristini, H. (2019). The new wave of luxury: The meaning and value of luxury to the contemporary consumer. *Qualitative Market Research: An International Journal*, *22*(3), 229–249.

Kim, Y. (2018). Power moderates the impact of desire for exclusivity on luxury experiential consumption. *Psychology & Marketing*, *35*(4), 283–293.

Klaus, P.P., Tarquini-Poli, A. (2022). Come fly with me: exploring the private aviation customer experience (PAX). *European Journal of Marketing*, *56*(4), 1126–1152.

Liu, H., & Li, X. (2021). How travel earns us bragging rights: A qualitative inquiry and conceptualization of travel bragging rights. *Journal of Travel Research*, *60*(8), 1635–1653.

Liu, H., Wu, L., & Li, X. (2019). Social media envy: How experience sharing on social networking sites drives millennials' aspirational tourism consumption. *Journal of travel research*, *58*(3), 355–369.

Loureiro, S. M. C. (2020). Virtual reality, augmented reality and tourism experience. *The Routledge handbook of tourism experience management and marketing*, 439–452. Routledge.

Martin, G., & Lewis, R. (2022). Gen X and the luxury dilemma: Legacy vs. sustainability. *Generation X Consumer Journal*, *11*(4), 28–39.

Michaelidou, N., Christodoulides, G., & Presi, C. (2022). Ultra-high-net-worth individuals: Self-presentation and luxury consumption on Instagram. *European Journal of Marketing*, *56*(4), 949–967.

Miller, R., & Rose, D. (2015). Luxury branding for Gen X: How to capture the elusive X-factor. *Brand Management Review*, *10*(3), 34–44.

Moscardo, G., & Benckendorff, P. (2010, November). Sustainable luxury: oxymoron or comfortable bedfellows. In *Proceedings of the 2010 international tourism conference on global sustainable tourism, Mbombela, Nelspruit, South Africa* (pp. 15–19).

Nueno, J. L., & Quelch, J. A. (1998). The mass marketing of luxury. *Business horizons*, *41*(6), 61–68.

Park, J. Y., & Lee, H. E. (2024). How Consumer Photo Reviews and Online Platform Types Influence Luxury Hotel Booking Intentions Through Envy. *Journal of Travel Research*. https://doi.org/10.1177/00472875241247317

Parker, S. (2020). Understanding Gen Z's approach to luxury. *Fashion and Culture Quarterly*, *8*(2), 54–61.

Peng, N., & Chen, A. (2019). Luxury hotels going green–the antecedents and consequences of consumer hesitation. *Journal of sustainable tourism 27*(9), 1374–1392.

Peters, N., & Lang, T. (2021). Socioeconomic indicators and luxury consumption. *Luxury Market Review*, *12*(3), 65–77.

Sakdiyakorn, M., Golubovskaya, M., & Solnet, D. (2021). Understanding Generation Z through collective consciousness: Impacts for hospitality work and employment. *International Journal of Hospitality Management*, 94, 102822.

Shahid, S., & Paul, J. (2022). Examining guests' experience in luxury hotels: Evidence from an emerging market. *Journal of Marketing Management*, *38*(13–14), 1278–1306.

Smith, J., & Andrews, L. (2019). Navigating the luxury market: Sustainability as the new norm. *Journal of Luxury Strategy*, *10*(1), 34–46.

Smith, J., & Luxmore, P. (2017). Luxury across generations: A comparative analysis. *Journal of Luxury Studies*, *5*(2), 7–21.

Smith, R., & Garcia, S. (2021). Tailored marketing strategies for luxury brands in a diverse market. *Branding and Strategy Quarterly*, *7*(1), 28–42.

Stevens, A. (2017). Gen X and the quest for authentic luxury. *Brand Insights Journal*, *9*(4), 48–56.

Sung, Y., Choi, S. M., Ahn, H., & Song, Y. A. (2015). Dimensions of luxury brand personality: Scale development and validation. *Psychology & Marketing*, *32*(1), 121–132.

Thomsen, T. U., Holmqvist, J., von Wallpach, S., Hemetsberger, A., & Belk, R. W. (2020). Conceptualizing unconventional luxury. *Journal of Business Research*, *116*, 441–445.

Tynan, C., McKechnie, S., & Chhuon, C. (2010). Co-creating value for luxury brands. *Journal of business research*, *63*(11), 1156–1163.

Vigneron, F., & Johnson, L. W. (1999). A review and a conceptual framework of prestige-seeking consumer behavior. *Academy of Marketing Science Review*, *1*(1), 1–15.

Vigneron, F., & Johnson, L. W. (2004). Measuring perceptions of brand luxury. *Journal of Brand Management*, *11*(6), 484–506.

Volo, S., & D'Acunto, D. (2021). Ecotourism as form of luxury consumption. In *Routledge handbook of ecotourism* (pp. 279–288). Routledge.

von Wallpach, S., Hemetsberger, A., Thomsen, T. U., & Belk, R. W. (2020). Moments of luxury–A qualitative account of the experiential essence of luxury. *Journal of Business Research*, *116*, 491–502.

Wiedmann, K. P., Hennigs, N., & Siebels, A. (2009). Value-based segmentation of luxury consumption behavior. *Psychology & Marketing*, *26*(7), 625–651.

Williams, L., & Green, R. (2022). Enrichment through travel: Beyond the luxury facade. *Global Travel Insights*, *13*(2), 23–35.

Yang, W., & Mattila, A. S. (2016). Why do we buy luxury experiences? Measuring value perceptions of luxury hospitality services. *International Journal of Contemporary Hospitality Management*, *28*(9), 1848–1867.

Yeoman, I., & McMahon-Beattie, U. (2018). The future of luxury: Mega drivers, new faces and scenarios. *Journal of Revenue and Pricing Management*, *17*, 204–217.

11

IMPACT OF HEALTH CRISIS (PANDEMIC) ON CONSUMER BUYING INTENTION

A lesson for future small businesses

Saloomeh Tabari and Wei Chen

Introduction

In December 2019, an unknown virus, linked to the wholesale seafood market in Wuhan, the capital of Hubei province, China, put the whole country and world at risk. 2019-nCoV, or COVID-19, is the seventh member of the family of coronaviruses and the third within the past two decades to emerge in the human population, after SARS-CoV in 2002, and MERS-CoV (the Middle Eastern Respiratory Syndrome) in 2012. It puts the World Health Organisation (WHO) on high alert (Zhu et. al. 2020; Munster et. al. 2020). Since December 2019, the illness has infected more than 621,365,294 people around the world (Aljazeera, 2020; WHO, 2020; 2022). The small businesses that drive the huge economy around the world are deeply concerned about how much damage this outbreak will cause to society.

Small businesses make a substantial socioeconomic contribution within the European Union (European Parliament, 2017) and shape Europe's business landscape (Statista, 2020). The main aim of this conceptual chapter is to reflect on previous epidemic diseases' effects and explore the potential impacts of COVID-19 on small businesses, and changes in consumer purchasing behaviour. Compared with large corporations, small businesses have less cash, resources and competence to cope with external crises. This COVID-19 outbreak is particularly unprecedented, with countries proposing total lockdowns. More noticeably, many small businesses have had to close. In the meantime, researchers have noticed that consumers have changed their consumption patterns during this crisis, such as a swift shift to online shopping for everything instead of going out to shops.

DOI: 10.4324/9781032637778-12

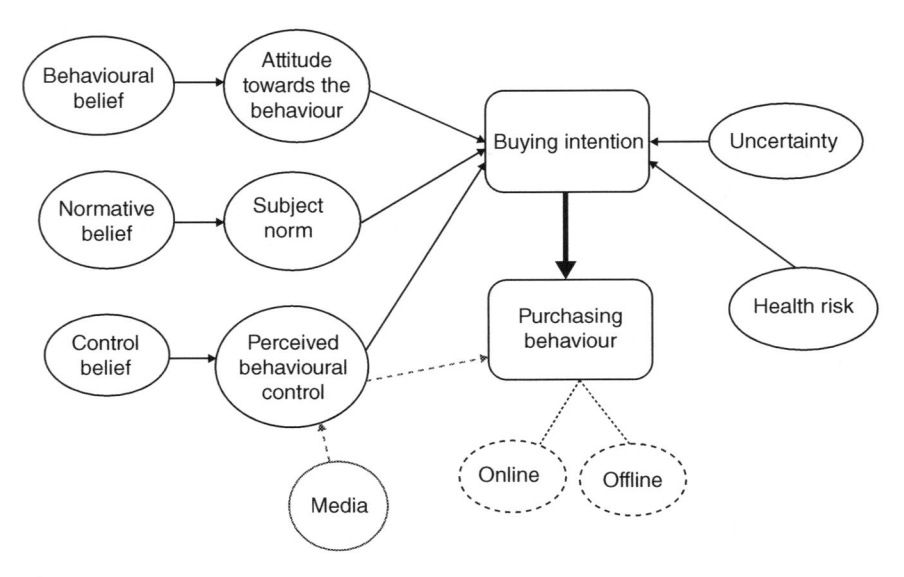

FIGURE 11.1 Conceptual framework: Impact of a health risk epidemic on purchasing behaviour.

Even though definitions and typologies of crises differ (Buchana and Denyer, 2011), the idea includes 'disasters, business interruptions, catastrophes, emergency or contingency' (Herbane, 2010: 46), which are low-probability events that exert a severe impact in terms of creating uncertainty and adding time pressure to decision-making (Pearson and Clair, 1998; Weick, 1988). COVID-19 can be considered the biggest crisis during the last decade and has had an immense effect on the world. The COVID-19 breaking required a comprehensive set of strategic measures and a combination of restrictions, including limited business-opening hours or the complete temporary shutdown of businesses, such as food providers (Qureshi et al., 2021). During a crisis, small businesses will face greater disruption because of the personal impact on the owner/manager and their greater vulnerability due to a lack of readiness, past experiences and resources (Runyan, 2006). Although the previous literature has discussed the barriers to the business development of entrepreneurs and small firms (Doern and Goss, 2013; Kouriloff, 2000; Lee and Cowling, 2013) and failure (e.g., Cardon et al., 2011; Shepherd, 2003), small business survival has not been examined from a crisis perspective (Herbane, 2010), especially during a health crisis and pandemic, with respect to changes in consumer behaviour. Moreover, a small number of studies within the crisis/disaster literature have examined their impact on small businesses (Wasileski et al. 2011; Rodríguez et al. 2004) or their responses (Runyan, 2006), but there is no framework to emphasise the changes on consumer behaviour.

TABLE 11.1 Summary of the length of the impact and changes with regard to the periods of the pandemic

Periods		Pre-start of pandemic	During-pandemic	Post-pandemic
Length of impact		Immediate	Medium/short-term	Long-term
Changes in buying behaviour of consumers and market	Buying	-Consumer: Irrational buying (panic buying) -Market: Not ready and supply change, distribution collapses	-Consumer: Slightly back to the rational buying but not the same as before pandemic -Market: Slightly getting ready with a high demand for an online-glossary and disruptions to the supply chain	-Consumer: Stock up on some essential products and the use of more delivery -Market: Adopting a new business model and providing more online facilities
	Resources	-Consumer: Using local resources (e.g., local supermarket, drugstores) -Market: Slightly relying on localisation	-Consumer: Local and regional resources and products (e.g., bigger, more central supermarkets, switch to local butchers and farmers) -Market: more focus on localisation and regionalisation	-Consumer: Stay with local and regional producers due to better price and better accessibility -Market: Anti-globalisation and reshaping the supply chain strategy to become more regional and national
	Behaviour	-Consumer: Heavily off-line buying as a result of panic buying and lack of readiness of the market -Market: Sudden shock to service, unprepared for a huge increase in demand	-Consumer: Combination of online and offline shopping with higher demand for onlineAdapting to digital and virtual life, less contact, cashless life -Market: Shortage of products, labour not ready for online and delivery processes	-Consumer: More online shopping, digitalisation and cashless (mobile pay) lifestyle -Market: Agreeing to change the old strategy to a digital and online strategy, change their business model, innovation and AI, mobile pay and non-physical banking

Furthermore, the crisis management literature has a tendency to focus on the 'association between planning and the enhancement of preventative actions and/or responses to organisational failures, accidents and interruptions' (Herbane, 2013: 83 cited by Doern, 2016). Previous studies mostly focused on large complex organisations such as emergency services in the context of natural disasters (e.g., McEntire et al. 2002; Quarantelli, 1988) or industrial crises (Buchanan and Denyer, 2012; cited by Doern, 2016), rather than health risk and pandemic on small businesses within service sector and change on consumer behaviour. Despite the global wide-ranging news about COVID-19, existing literature on entrepreneurship and small businesses had a limited focus on this topic. In other words, managers and practitioners have no reference regarding how to deal with the economic crisis in the country and their business field during a pandemic. The current situation is not unprecedented and, unfortunately, will not be the last crisis to arise. The existing crisis management literature has not dealt in depth with small businesses' responses to health disasters, so further research is required. This conceptual chapter aims to raise awareness of the issue for the industry and managers by proposing a conceptual framework for consumer purchasing behaviour during a period of health crisis and uncertainty. As such, this chapter proposes a conceptual framework to understand these changes during the period of health risk and uncertainty. The framework enables us to understand consumer reactions and their purchasing behaviour during a health disease disaster compared to other disasters. The chapter provides five propositions for consideration in the future study. Could the coronavirus be the straw that breaks the camel's back? It is unclear how many small companies will ultimately feel the full impact of the virus, survive or even benefit from opportunities to revise or change their business model.

Lessons from the past and future

Epidemics and pandemics do not just come and go, they have an impact on society and economy. For instance, the epidemic in early 1832 hit France and other European cities hard (O'Sullivan, 2021). The Spanish Flu in 1918 affected most of Europe and the USA. With the enforcement of various restrictions across cities and countries, businesses were forced to shut down because so many employees were sick (History.Com, 2020). A few researchers showed that the Spanish flu pandemic gave new way to businesses, with start-ups booming from year 1919 as a result of the pandemic (Beach et al., 2020; Karlsson et al., 2014).

A small number of studies have explored the factors that affect business survival and recovery, from pre-crisis characteristics and planning to the nature and extent of the damage and response to the damage during natural disasters, like hurricanes, earthquakes or floods (Rittichainuwata & Chakraborty, 2012) or crisis situations (Gummesson, 2009; Schenker-Wicki et al., 2010). Furthermore, Corey and Deitch (2011) argue that, despite the fact that 'studies have attempted to determine features that predict the success of recovery for individual businesses post-disaster the results have been inconsistent regarding what matters most' (p.170). Previous studies failed to consider factors like health risks and pandemics during crises and their impact on consumer purchasing behaviour with regard to small businesses. Thus, it remains unclear how these small businesses need to act.

After nearly two years of COVID-19 and a second lockdown in some parts of China on 28 March 2022 (news.sky.com, 2022), it seems we are still at an early stage to talk about the pandemic on business and economy around the world, but so far, it has had a huge impact on the trade and travel industry all around the world and forced lots of businesses to change their strategies and business models. Seetharaman (2020:3) argues that

> some may say that these shifts are temporary reactions to the pandemic and once routine is back, businesses will go back to their previous business model or find a new stability to settle at. However, the pandemic has offered an opportunity to digitize a business or identify an alternative business model.

Although the Chinese government acted very quickly, closing off the source of the virus (the seafood market) and putting people in isolation, the virus swiftly travelled to other cities and countries. Many shops were closed across China and lots of airlines cancelled flights to and from China, such as Air Canada (CNBC, 2020). Economic turmoil and health risks have forced a large proportion of the service sector and retail shops to close in mainland China and lots of factories remained closed even after the Chinese New Year celebrations. As a result, economists and analysts forecast a decline in China's GDP as the virus hits the world's second-largest economy (CNBC, 2020). The COVID-19 outbreak has already changed the schedule of all international and domestic travellers who plan to travel to Mainland China, Hong Kong, Singapore, Malaysia, Vietnam, South Korea and Japan. Many countries have followed the same procedure as was in place during SARS and have issued travel advisory and health alert notices to their residents (WHO, 2022).

Furthermore, as we can see from previous outbreaks of similar viruses, such as Ebola, SARS and influenza, these had an enormous impact on society and panic spread around the globe, with trust declining towards the people and areas associated with hosting the virus. However, the current COVID-19 outbreak has affected not only China but also other countries like Hong Kong, the Middle East, Europe and other Asian countries. In similar situations in the past, fear and a sense of insecurity, locally and internationally, grew and, because of the nature of the virus, which is transferred from human to human, people isolate themselves and avoid daily activities, such as visiting local restaurants or cafes, cinemas, shopping malls and even using public transport. As a result, many small and family businesses within the service sector will suffer. The questions are: 'when they will be able to start work again?' and 'when will people feel confident about eating outside the home and buying seafood and meat again?' In addition, many countries have stopped importing goods from China; for example, Indonesia has stopped all fresh fish/seafood imports from China, with the Indonesian Ministry of Maritime Affairs and Fisheries commenting: 'We were asked temporarily to stop the imports. We do not want to take a high risk, too' (SeafoodSource, 2020).

Ingram et al. (2013) highlighted that the results of actual medical epidemics, natural disasters and political events can be represented by tourist receipts in countries; for example, the SARS epidemic of 2002 and the Tsunami of 2004 reduced inbound tourism to Thailand. Thai tourism flattened between 2006 and 2010, perhaps not only due to the well-publicised riots but also due to the world recession that occurred during that period. Since that time, tourism has grown considerably. They have suggested that tourism responds immediately to actual terrorism, but it is difficult to isolate terrorism as the sole cause, as the worldwide economic downturn might also have been a contributing factor. Regarding COVID-19 and its impact on the economy of China and the rest of the world, it would be difficult for the tourist industry, especially the service sector and small businesses, to recover quickly.

The panic and fear among people from different countries, such as South Korea, Malaysia and, more recently, Italy and Iran, have exacerbated the concern. When the number of cases outside China reached 1,200, the virus changed from an epidemic to a pandemic (BBC, 2020). The uncertainty regarding the situation with the COVID-19 disease to date has caused people to react dramatically and has had a direct impact on their intentions and perceived behaviour. The sudden demand for food and preparation for isolation has caused economic instability (such as in the northern part of Italy). Moe and Pathranarakul (2006) mentioned that disaster management is used alongside emergency management, which involves plans, structures and a coordinated response to the whole scale of emergency needs which should be made urgently when a disaster occurs. Previous research argued that small businesses are not effectively prepared for crises and lack the resources to fulfil what is required in such a situation. According to several managers, their day-to-day problems are so difficult and time-consuming that they do not have enough time to plan for upcoming crises and uncertainty (Barton, 1993; Caponigro, 2000).

Normal life, social interactions and businesses have been interrupted since people have been told to avoid crowded places and not eat any raw food. Lots of grocery shops and restaurants, as independent businesses, in the small business category, have had to close based on the subject norm. Zhang et al. (2005) highlighted the result of research on SARS by Feng (2003), that consumption is directly related to psychological factors during an epidemic. The report shows that COVID-19 has had a huge impact on the service and restaurant industries, such as well-known Chinese chain restaurants such as Xibei Oat Noodle Village and Home Original Chicken. Furthermore, in other sections, Foreign-invested enterprises have also been affected; companies like Tesla and Apple have closed their outlets in China, which has had a direct impact on the buying behaviour of consumers (Mao and Zhang, 2020; He, 2020).

What we have learnt, and what should we do?

Due to the isolation policy and fear among the general public, many retail shops and the service sector have been affected alongside small businesses and entrepreneurs. Diseases have a longer impact compared to natural disasters, terrorism and political issues, which makes it very difficult for small and family businesses to return to normal. What can they do to limit the damage?

Limited research has focused on the impact of health risks on small and family businesses and the changes in buying behaviour. Due to the nature of the service industry, most of the components, such as grocery shops, cafes, restaurants and small factories, are considered small businesses and entrepreneurs, some of which are run by family members. Previous studies argue that, during crises, no business is immune from devastation and those small businesses and entrepreneurs, with their limited resources, are more vulnerable to the catastrophe and unprepared for unexpected crises. COVID-19 put a large strain on entrepreneurs and small businesses, who experienced an unprecedented shock to their businesses (Torres et al., 2021). The results of research by Spillan and Hough (2003) confirmed that small businesses have little concern about crises that they have not experienced before and having a crisis management team is of little assistance, as their focus is on past experiences. Fok et al. (2019) highlighted that disaster management includes different time paths for each disaster with variable impacts. Therefore, small businesses are unprepared for pandemics or sudden disasters, which cause a direct change in consumers' attitudes and purchasing behaviour, such as compulsive, irrational or panic buying.

Only a handful of studies have focused on health crises/outbreaks (e.g., Foot and Mouth disease, Ebola, swine flu and influenza) and their impact on small businesses, and no research has specifically assessed health risks and consumers' reactions to these with regard to small businesses. Irvine and Anderson (2004) focused on small rural businesses operating in the UK tourist industry to examine the impact of the Foot and Mouth disease outbreak. Their results highlight that the effects were significant, with major reductions in the volume of business, staff numbers and profitability, particularly for micro-businesses. Many closed, some took up to a year to recover, while others reduced their expenditure and created alternative ways of generating profits. The result of another study by Runyan (2006) regarding the impact of a natural disaster on small businesses, in this case the effect of Hurricane Katrina in the United States, found evidence of four conditions of a crisis – low probability, ambiguity, high consequence and decision-making time pressure (Pearson and Clair, 1998). Several barriers to recovery were also identified, including access to capital, diminished cash flow, poor communications and disruption to services (e.g., electricity). In addition, environmental uncertainty plays an important role and often means fewer business opportunities (Rastkhiz et al., 2018). Tabari and Chen's (2022) findings showed that the female entrepreneur's potential innovative and determined behaviour were the factors in the resilience of their businesses and created various cycles of learning, anticipation and adaptation, which resulted in tapping into the new market.

It is essential to examine the attitudes and behaviour of individuals and communities with regard to small businesses. Doern (2016) argued that small businesses can be more resilient and less vulnerable to crises by drawing on and adapting to their past experience, developing a mindset that is both anticipation- and containment-oriented and investing in and building resources. This cannot happen without developing a clear picture of past experiences and a framework for evaluating the situation. Walker and Page (2003) stated that risk reception is subject to variable intensity and severity levels, like terrorism, which is rare, but its consequences are severe, while health risks generally have a high frequency. In such a situation, when health risks threaten the country as a whole, the behavioural beliefs and attitudes towards the behaviour have a direct impact on people's consumption behaviour. Therefore, the researchers adopted the Theory of Planning Behaviour (TPB) and proposed a conceptual framework for building trust within the community and customers to help them and provide greater community support. Risk management planning is important in enabling small and family businesses to understand the elements of risk management and how to implement it in situations of this nature.

The media plays an important role during all disasters, from terrorism to any epidemic or pandemic like COVID-19, but the question arises as to how to handle the media. The media are not always a source of truth and, at times, one of the main resources for providing false and incorrect information, which creates even greater fear among the community. However, businesses have reported that, during the SARS outbreak, the number of online customers and online orders increased rapidly, so SARS convinced millions of people, who were too afraid to go out and remained in isolation, to try shopping online. This change in consumption behaviour was inevitable; for example, Alibaba changed from B2B to B2C with the help of Taoba and built its e-commerce (qz.com, 2020 and China-briefing.com, 2020). E-commerce and online shopping changed consumer shopping behaviour during the 2003 SARS epidemic. Ordering groceries and high-quality takeaways have become extremely popular during the last few months in China because of COVID-19, meaning that food e-commerce (online order for delivery) is witnessing high growth, which can provide an opportunity for small businesses to offer online shopping facilities and update their status on social media to gain trust, but businesses should remember not to push coronavirus messages in their communication and marketing and try to focus more on philanthropic initiatives. Online grocery shopping increased rapidly during a ten-day period of COVID-19, from late January to early February, according to Carrefour and JD.com (Cheung, 2020). Laskovaia et al. (2019:5) addressed that 'during the time of economic crisis, businesses with higher levels of entrepreneurial orientation are more likely to take advantage of new business opportunities'. In this vein, small businesses with a higher level of willingness to be more entrepreneurial orientations are able to use the changes within environments as an opportunity for renovation to better survive in the challenging time. This disruptive crisis needs to be managed more positively (Ratten, 2020). Prokop et al. (2024) stated that the Chinese migrant entrepreneurs in the service sectors responded to the crisis with two approaches of adaptation and hibernation.

Tsinghua University and Peking University have researched on 1000 small- and medium-sized businesses which claimed to be able to survive for a month if they continued to keep their business closed and the nature of their business did not provide the flexibility for their employees to work from home. The contribution of entrepreneurs to small- and medium-sized businesses is high in today's market. For instance, China alone has roughly 30 million small- and medium-sized businesses, which contribute 60% of the GDP of the country (He, 2020).

Theory of planned behaviour

The theory of planned behaviour (TPB) originally comes from social psychology research and works on the assumption that intention is a significant predictor of behaviour, while intention itself is a function of behavioural beliefs that link the given behaviour to certain outcomes or actions (Ajzen 1988, 1991; Kautonen et. al. 2011; Linan and Chen, 2009). TPB has been used as an expectancy-value model of attitude–behaviour relationships which details the determinants of an individual's decision to endorse a particular behaviour and is designed to provide parsimonious explanations of informational and motivational influences on behaviour (Conner and Armitage 1998). In other words, TPB assumes that individuals make decisions based on a careful consideration of the known available information, and it specifically addresses behaviour or situations where individuals have inadequate ability or control to employ their own determination (Ajzen 1988, 1991, 2002).

TPB elements are considered individual intention (INT) and perceived behavioural control (PBC). Intention is detained to be the motivational element that urges an individual to engage in a particular behaviour which is acceptable or predictable by society. PBC represents the extent to which people have control over engaging in the behaviour. In succession, intentions are determined by an individual's attitude towards the behaviour (ATB), subject norms (SN) and perceived behavioural control (PBC), which are referred to as the direct predictors. Different sets of beliefs and unpredicted situations trigger ATB, SN and PBC and are referred to as indirect predictors. With regard to the model, attitudes are involved with beliefs about the likelihood of noticeable outcomes of the behaviour being prejudiced by the evaluation of each outcome. Subjective norms are linked to beliefs about whether noticeable referents think that one should perform the behaviour prejudiced by the motivation to fulfil with that referent. PBC is involved with beliefs about the frequency of the occurrence of facilitating or inhibiting factors towards engaging in behaviour subjective by the perceived power of each element to impact engagement with the behaviour (Conner and Sparks, 2005).

The TPB model has been used in the entrepreneurial context and so contributes to the existing literature on the emergence of entrepreneurial behaviour prior to the onset of any observable action, which has distinguished implications for policy (Kautonen et. al. 2009; Linan and Chen, 2009). Researchers like Johnston and Dixon (2008), McEachan et al. (2010) and Michie et al. (2005) have proposed that the nature of these similarities or differences amongst types of behaviour may be important moderators of the ease with which we can predict or change them. Although TPB has been criticised by some researchers for failing to provide specific guidance on change techniques, its providence makes the model generally useful in applied settings. For the purposes of the current research, health-risks, here COVID-19 and uncertainty behaviour were defined as behaviour which impacts or has the potential to impact upon the health of an individual and society in a negative way and will have an impact on buying intention and so a direct impact on purchasing behaviour, leading to two types of purchase, online and offline. Furthermore, during any type of epidemic and pandemic, the media play a vital role as a moderating factor on perceived behavioural control and purchasing behaviour. As we have seen during the past few months, the pandemic has changed consumer purchasing habits in different countries and the authors believe that factors such as health-risks and uncertainty lead to irrational and panic buying, in terms of offline and online and some has a medium/short-term and some long-term effect.

Our application of TPB as a perspective on the impact of a health risk epidemic on purchasing behaviour is justified accordingly. First, it identifies the central link between intention and behaviour, which reflects the fact that people tend to engage in behaviour that they intend to perform. In the case of an epidemic, buying intention will be reflected by two types of purchasing behaviour (online and offline) and the purchase will be a strategic choice that is actively followed. Second, the link between PBC and behaviour shows that individuals are more likely to engage in the required behaviours over which they have control and will be prevented from enacting behaviour over which they have no control. During an epidemic, however, the media, as a moderator, influences people's decisions alongside health risks and uncertainty as catalyst factors. As discussed earlier, the three sets of beliefs (behavioural, normative and control) can be triggered by even very minor changes in an individual's environment. Therefore, the current crisis as a key tool is a very powerful trigger for these three sets of elements. Health risk and uncertainty about the future act as catalysts during the current catastrophe. Consequently, the trigger for such elements, the uncertainty and health-risk lead to an unknown future and lack of trust in the current pandemic which provides a faster reaction in terms of an individual's attitude, subject norm and perceived behaviour and leads to different lengths of reactions: immediate (very short-term), short-term and long-term. According to TPB theory, these changes should be predictable in normal circumstances but, in the case of a health-risk threat, some of the reactions are unpredictable, like consumers' buying intentions. The health-risk element has played an important role during the current global situation regarding the purchasing behaviour of consumers with a terrifying reaction to their behaviour which led to a market shock; for instance, their irrational or panic buying of toilet paper, which started in Australia, food purchases (such as pasta and flour) and food waste in the UK and gun purchases in the USA. The proposed conceptual framework aims to help small businesses and the market understand the changes to customers' purchasing behaviour, by adding health-risk and uncertainty as catalysts and the media as a moderator to the TPB concept and highlighting the importance of disease management as an element in the crisis/risk management literature. As such, the conceptual framework will help small businesses to improve their preparedness for crises, despite their lack of experience and readiness to face disasters like the current pandemic.

Discussion

The current situation can be named a novel situation in the history of humankind in the recent century after the Great Depression of the 1930s and the Great Recession in the 2000s. The Great Lockdown (2019–2020) happened at a most unexpected time, with no one prepared for economic turmoil. The Great Recession prepared the market for economic disruption, with people gradually changing their behaviour and they did not close the whole market overnight. On the other hand, the great lockdown happened suddenly with great uncertainty and concern around the world. The pandemic affected some countries severely and only a few nations were able to exercise a higher degree of control over the situation. To date, a few airlines announced their bankruptcy, and many other businesses are in a grey area of an unknown future.

Fok et al. (2019) mentioned that more significant and long-lasting effects on the market can be seen from classical market disasters. However, the current dynamic situation of the market is not comparable to classical disasters, as the effect of the pandemic will have a stronger impact on the market, in particular from the variable duration of the economic and behavioural aspects. Nearly five months after the first case was reported by China, little is still known about COVID-19 and this brings more uncertainty to businesses, especially small firms, which do not have enough power to bargain with their suppliers.

Moreover, the pandemic caused a simultaneous shock to the economy, to both the demand and supply channels. The change in purchasing behaviour and individual attitude gave a shock to the market demand, which led to a shortage of some hygiene and food products. Ivanov (2020) stressed that a sudden disruption of demand leads to negative impacts on supply chain performance. Furthermore, based on changes in purchasing behaviour, disruption in the supply chain, a sudden fall in global trade, the halting of global exports and imports and changes in market demand, new habits will shift the performance of the market; nonetheless, some of these changes will be medium- or short-term adaption to behaviour and a few will be long-term adoption.

Based on the proposed conceptual framework, we believe that a health-risk epidemic, and especially the current pandemic, can be divided into three different periods. *The first period* is the 'pre-start of the pandemic', when the market was not ready and consumers showed unpredictable reactions, which were mostly immediate reactions and stayed for a very short-term, such as offline panic buying, and very quickly it moved on to the next period. *The second period* is the 'during-pandemic', when most countries went into lockdown, the global market felt a sudden shock and shift on demand, rational behaviour and adapting to the new normal, showing more flexibility, more contactless, cashless, online shopping, digitalisation and Artificial Intelligence (AI), Robot (nurses, delivery) and looking for local products, resources, helping the local community and regionalisation. In this period, individuals have shown great flexibility towards the 'new normal' with more rational behaviour. They have adapted quickly and made changes to their daily life, such as working from home, by embracing digital technology, switching to alternatives, using online gyms and buying gym equipment for home that follows 'physical distancing' but, at the same time, is more 'socially active' through digital platforms such as Hang out, Zoom, Skype, WhatsApp, Houseparty and many more. Even senior citizens using online platforms as well as virtual connections indicate a broad age characteristic of the group; for example, an online retailer in China during COVID-19 reported a 237% increase in users over 40 years old (Nielsen.com, 2020). These individuals will retain some of these changes even after the pandemic, instituting long-term shifts in their purchasing behaviour and digitalisation. As Seetharaman (2020) mentions, still there is a gap concerning understanding the changes to digitalisation and business model changes to respond to COVID-19 and changes in behaviour. The result of a study by Penco et al. (2022) articulated the fact that during COVID-19 all of the companies in their study invested in some type of web-based technologies, like social media, websites and e-commerce to develop and maintain their customer relations. However, as mentioned earlier not all small businesses can afford this investment. The pandemic brought some unprecedented changes to people's lifestyles and behaviour and also to businesses regardless of their size and nature of business.

Some of these changes will have a more long-term effect and will be adopted in future to reshape the business model and market and, as a result of that, a new economic equilibrium may arise for the next period. *The third period* is the 'post-pandemic' era, which is more about recovery and survival time for the market and businesses, as they need to be aware of the new behaviours and changes adopted by the consumers, because these changes and shifts in the market will be inevitable and, to survive in the future, they need to accept the changes. The long-term effects will be seen in the market, such as the importance of hygiene equal to security, and online-shopping will be more and more in demand. Before the pandemic, the retail market and high street shops were facing financial difficulties, since customers, day by day, were switching to more online-shopping, especially the younger generation, as a result of which many of them have closed their high street branches and COVID-19 can be the last push for big mega stores and the retail industry. Economic shifts from global trade to more localised sources could be one of the long-term changes that the market could face as a healing solution regarding the domestic market and for small/local businesses, like buying from local butchers and domestic travelling. Accepting AI in the service industry more and more, there are more robot doctors, nurses, robot hotel receptionists and robot delivery as we move into a more digital phase, like using smart mirrors to try on clothes during online shopping, digital/virtual gyms and more distance working from home and out of offices, the use of more virtual meetings, and attending events (weddings, funerals, birthday parties) more virtually instead of travelling to the places to attend in person. We have highlighted some of the behavioural changes in the different periods in the market and the new shifts in the market. With regard to the market shifts (Table 11.1), proposed conceptual framework and the provided discussion on the impact of different periods on behavioural changes, we suggest the following propositions:

PROPOSITION 1: Uncertainty and health risks will trigger buying intentions that lead to different lengths (immediate, medium and long-term) of changes regarding buying intention.

PROPOSITION 2: A health-risk epidemic will have an impact on the purchasing behaviour of individuals.

PROPOSITION 3: Purchasing behaviour in the pre-start era of the pandemic will be immediate, more irrational and off-line.

PROPOSITION 4: Purchasing behaviour during the pandemic will move to more online, local and regional products.

Conclusion

The current situation regarding COVID-19 remains unsettling and the fear is growing day by day in different countries, not only China, the source of the virus, since the status of the virus has changed from an epidemic to a pandemic and almost 228 countries, at the time of writing, have been affected (Worldometers, 2022). The global economy is declining, which will lead to a recession after the pandemic. Due to a fear of crowds and the need for physical distancing, countless people have cancelled their flights and many airlines no longer fly or over very limited flights to affected countries. The global catastrophe has changed the world, has been referred to as a war experienced by countries dealing with it due to the sudden halting of global trade and a lack of essential products on the market and has provided an unfamiliar situation to the post-war generation, who have never lived through such a disruptive event before. Even if we embrace the COVID-19 crisis within the next few months, the damage caused by this pandemic to people's life and wellbeing as well as the market will continue to live in our memories for many years to come.

The health risk and uncertainty have affected people's lives and shifted the market demand and consumer purchasing behaviour. The service, catering, tourist and hospitality industries have been affected in most countries, with the highest number of cases. These industries have a direct impact on the economy of the countries and many small businesses will be affected, due to the halting of trade and collapse of the stock market. The global recession will be felt sooner or later as a result of the pandemic, as the situation was unpredictable, and the market has not experienced a global economic crisis resulting from a public health emergency before, not even during the Spanish flu outbreak. It is uncharted territory; more studies need to explore these changes and its impact on business and the economy since the threat of the virus remains high and unknown. The lack of readiness of the market made the current situation very expensive experience for the market. In total, lockdown, mobility restriction (physical distancing) and travel bans have put the service industry and small businesses more under pressure and a lack of huge savings have made their healing time and survival unclear. A lack of time for preparation was another issue that small businesses dealt with, and the sudden order from governments for lockdown as well as a lack of past experiences made it more difficult for them.

Shifting people's behaviour towards e-commerce, such as buying fresh food, groceries, entertainment, education and takeaways online, will be the main factor that small businesses can consider. Providing contactless products, without any human-to-human interaction, may be one of the factors that restaurants, cafes, grocery shops and retail outlets can offer, plus offering packaging that is free from contamination. Meanwhile, the data show that the usage of the internet, TV and applications such as Netflix has grown during the isolation period, since people cannot leave their homes (Cheung, 2020). According to JPMorgan Chase's Joseph Lupton, China's economy will eventually bounce back from the coronavirus and grow by 15% in the second quarter on a quarter-on-quarter, annualised basis, with a potential 4% growth in the first quarter (Stankiwicz, 2020). Investors are grappling with the potential global economic effects of the virus, which has disrupted the supply chains in many countries as factories shut down and workers stay at home.

Even the health sector has benefited greatly from online platforms, according to Mao and Zhang (2020) since, during the COVID-19 epidemic in China, healthcare platforms such as Ali Health and Tencent HealthCare have provided free services and helped to distinguish cases from those with a common cold. This marks the beginning of telemedicine demand and will help the health sector to use the technology, based on the changes in people's consumption behaviour. The conceptual framework proposed here could help small businesses and even the retail sector to understand the changes to purchasing behaviour and the importance of the media as a moderator in this regard. As mentioned before, health risks and health disasters have a longer-term impact on people's behaviour compared to natural disasters or terrorism, and their effect and time scale are predictable, but the SARS lesson showed that health risks have a longer-term impact on the image of the country and the economy. However, with the limited facts available about COVID-19, this virus appears stronger and to have affected more countries compared to previous ones. Therefore, small businesses may use the proposed framework and provide online shopping systems as a strategy to help them to survive during the current economic turbulence by adapting to the changes in people's consumption behaviour and life conditions. Small businesses, because of their lack of experience and resources, will be affected more by any type of crisis and disaster, so they need to consider having a crisis management team in place to think beyond today's problems and make them to fight back better during the crisis. It is possible to recognise these traces of 'impulse behaviour in economic crisis' (Boutsouki, 2019:974) and analyse the changes on trends and behaviour among different sectors and perhaps they 'can survive and even grow so that the whole economy can benefit from higher level of resilience' (Cannavale et al., 2020:1005).

Furthermore, the proposed conceptual framework and research propositions can be used to inform crisis management during a health-risk pandemic and disease epidemic and how businesses could prepare for pre-start, during and post-epidemic. Any future outbreak of disease may be a reminder of the COVID-19 pandemic to consumers. Future studies can look at each country as a distinct case, and this study did not look at a specific country, to understand consumer purchasing behaviour by applying a proposed conceptual framework to find the consumer reaction and buying intention as a nation. We suggest adding 'trust in the government' as another element to their study. Future studies should consider the age group as another factor linked to buying intention that has a direct impact on individuals' attitudes during a health-risk crisis.

References

Ajzen, I. (1988). *Attitudes, personality and behaviour*. Open University Press.

Ajzen, I. (1991). The theory of planned behaviour, *Organizational Behaviour and Human Decision Processes*, 50, 179.

Ajzen, I. (2002). Perceived behavioural control, self-efficacy, locus of control, and the theory of planned behaviour, *Journal of Applied Social Psychology*, 32(4), 665–683.

Aljazeera. https://www.aljazeera.com/news/2020/02/cloneofcloneofcloneof200215224437270-20021623180-200218231013395.html, Accessed on 19/02/2020.

Barton, L. (1993). *Crisis in organizations: Managing and communication in the heat of chaos*. South-Western Publishing.

Beach, B., Clay, K., & Saavedra, M. H. (2020). The 1918 influenza pandemic and its lessons for COVID-19. National Bureau of Economic Research Working Paper Series, (w27673).

Buchana, D.A., & Denyer, D. (2011). Research tomorrow's crisis: methodological innovations and wider implications, *International Journal of Management Reviews*, 15, 205–224.

Boutsouki, C. (2019). Impulse behavior in economic crisis: a data driven market segmentation. *International Journal of Retail and Distribution Management*, 47(9), 974–996.

Cannavale, C., Zohoorian Nadali, I., & Esempio, A. (2020). Entrepreneurial orientation and firm performance in a sanctioned economy – Does the CEO play a role? *Journal of Small Business and Enterprise Development*, 27(6), 1005–1027.

Caponigro, J. R. (2000). *The Crisis Counsellor: A Step-by-Step Guide to Managing a Business Crisis*, Contemporary Books.

Cheung, M.C. (2020). Coronavirus' impact on consumer and businesses in China. https://www.emarketer.com/content/coronavirus-china-us-covid-19-impact-retail-travel, 6th February, Accessed on 28/02/20.

China internet boost. https://qz.com/662110/chinas-internet-got-a-strange-and-lasting-boost-from-the-sars-epidemic/, Accessed on 28/02/2020.

CNBS. https://www.cnbc.com/2020/02/05/china-gdp-2020-banks-trim-forecasts-amid-outbreak.html, Accessed on 28/02/2020.

Conner, M., & Armitage, C. J. (1998). Extending the theory of planned behaviour: A review and avenues for further research, *Journal of Applied Social Psychology*, 18, 1429–1464.

Conner, M., & Sparks, P. (2005). Theory of planned behaviour and health behaviour. In M. Conner & P. Norman, *Predicting health behaviour*, 2nd ed., pp. 170–222, Open University Press.

Coronavirus: World must prepare for pandemic, says WHO, https://www.bbc.co.uk/news/world-51611422, Accessed on 26/02/2020.

Doern, R. (2016). Entrepreneurship and crisis management: the experiences of small businesses during the London 2011 riots. *International Small Business Journal*, 34(3), 276–302.

Doern, R. & Goss, D. (2013). From barriers to barring: Why emotion matters to entrepreneurial development, *International Small Business Journal*, 31(5), 496–519.

Fok, D., Stel, A.V., Burke, A., & Thurik, Roy. (2019). How entry crowds and grows markets: The gradual disaster management view of market dynamic in the retail industry, *Annals of Operations Research*, 283, 1111–1138.

Gummesson, E. (2009). The global crisis and the marketing scholar. *Journal of Customer Behaviour*, 8(2), 119–135.

He, L. (2020). Small businesses drive China's economy, the coronavirus outbreak could be fatal for many. https://edition.cnn.com/2020/02/14/economy/coronavirus-china-economy-small-businesses/index.html, 14th February, Accessed on 28/02/22.

History.com (2020). Spanish Flu. https://www.history.com/topics/world-war-i/1918-flu-pandemic. Accessed 11/04/2022.

Worldometer (2022). https://www.worldometers.info/coronavirus/, Accessed on 11/04/2022.

Ingram, H., Tabari, S. & Watthanakhomprathip, W. (2013). The impact of political instability on tourism: case of Thailand. *Worldwide Hospitality and Tourism Themes*, 5(1), 92–103.

Irvine, W, & Anderson, A. (2004). Small tourist firms in rural areas: Agility, vulnerability and survival in the face of crisis, *International Journal of Entrepreneurial Behaviour and Research*, 10(4), 229–246.

Ivanov, D. (2020). Predicting the impacts of epidemic outbreaks on global supply chains: A simulation-based analysis on the coronavirus outbreak (COVID-19/SARS-CoV-2) case. *Transportation Research Part E: Logistics and Transportation Review*, 136.

Johnston, M. & Dixon, D. (2008). Current issues and new directions in psychology and health: What happened to behaviour in the decade of behaviour?, *Psychology & Health*, 23, 509–513.

Karlsson, M., Nilsson, T., & Pichler, S. (2014). The impact of the 1918 Spanish flu epidemic on economic performance in Sweden: An investigation into the consequences of an extraordinary mortality shock. *Journal of Health Economics*, 36, 1–19.

Kautonen, T., Tornikoski, E.T. & Kiber, E. (2011). Entrepreneurial intentions in the third age: The impact of perceived age norms. *Small business Economy*, 37, 219–234.

Laskovaia, A., Marino, L., Shirokova, G. & Wales, W. (2019). Expect the unexpected: examining the shaping role of entrepreneurial orientation on causal and effectual decision-making logic during economic crisis. *Entrepreneurship and Regional Development*, 31(5–6), 456–475.

Linan, F. & Chen, Y.W. (2009). Development and cross-cultural application of a specific instrument to measure entrepreneurial intentions. *Entrepreneurship Theory and Practice*, 33, 593–617.

Liu, F. (2003). Analysis and Solutions of the Impact of SARS on China's Tourism Industry, Special Issues on SARS 11, National Report.

Mao, V. & Zhang, B. (2020). New business opportunities emerging in China under Covid-19 outbreak. https://www.china-briefing.com/news/china-business-opportunities-covid-19-outbreak/, 18th February, Accessed on 28/02/20.

McEachan, R.R.C., Lawton, R.J. & Conner, M. (2010). Classifying health behaviours: Exploring similarities and differences amongst behaviour. *British Journal of Health Psychology*, 15, 347–366.

Michie, S., Johnston, M., Abraham, C., Lawton, R., Parker, D. & Walker, A. (2005). Making psychological theory useful for implementing evidence based practice: A consensus approach, *Quality & Safety in Health Care*, 14, 26–33.

Moe, T.L., & Pathranarakul, P. (2006). An integrated approach to natural disaster management. *Disaster Prevention and Management*, 15(3), 396–413.

Pearson, C.M. & Clair, J.A. (1998). Reframing crisis management, *Academy of Management Review*, 23, 59–76.

Penco, L., Profumo, G., Serravalle, F. & Viassone, M. (2022). Has COVID-19 pushed digitalisation in SMEs? The role of entrepreneurial orientation. *Journal of Small Business and Enterprise Development*. https://doi.org/10.1108/JSBED-10-2021-0423

Prokop, D., Tabari, S., & Chen, W. (2024). Survival instincts of Chinese entrepreneurs in the UK: adaptation or hibernation. *International Journal Entrepreneurship and Small Business*, https://doi.org/10.1504/ijesb.2025.10059695.

Qureshi, A. I., Suri, M. F. K., Chu, H., Suri, H. K., & Surid, A. K. (2021). Early mandated social distancing is a strong predictor of reduction in peak daily new COVID-19 cases. *Public Health*, *190*, 106–167. https://doi.org/10.1016/j.puhe.2020.10.015

Rastkhiz, S. E. A., Dehkordi, A. M., Farsi, J. Y., & Azar, A. (2018). A new approach to evaluating entrepreneurial opportunities. *Journal of Small Business and Enterprise Development*, 26(1), 67–84.

Ratten, V. (2020). Coronavirus disease (COVID-19) and sport entrepreneurship. *International Journal of Entrepreneurial Behavior and Research*, 26(6), 1379–1388.

Rittichainuwata, B. N., & Chakraborty, G. (2012). Perceptions of importance and what safety is enough. *Journal of Business Research*, 65(1), 42–50.

Runyan, R.C. (2006). Small business in the face of crisis: Identifying barriers to recovery from a natural disaster. *Journal of Contingencies and Crisis Management*, 14, 12–26.

Schenker-Wicki, A., Ianunen, M., & Olivares, M. (2010). Unmastered risks: From crisis to catastrophe: An economic and management insights. *Journal of Business Research*, 63(4), 337–346.

SeafoodSource. https://www.seafoodsource.com/news/supply-trade/coronavirus-concern-has-indonesia-restricting-imports-of-live-fish-from-china, Accessed on 19/02/2020.

Seetharaman, P. (2020). Business models shifts: impact of Covid-19. *International Journal of Information Management*, 54, 102173.

Spillan, J. & Hough, M. (2003). Crisis planning in small businesses: Importance, impetus and indifference, *European Management Journal*, 21 (3), 398–407.

Sky News (2022). https://news.sky.com/story/covid-shanghai-reports-record-cases-amid-unrest-over-lockdown-rules-but-china-stands-by-policy-12587809, Accessed on 13/04/2022.

Tabari, S., & Chen, W. (2022). Ethnic female entrepreneurs in the service sector: Challenges and motivations. In S. Tabari, & W. Chen (Eds.), *Global strategic management in the service industry: A perspective of the new era.* Emerald Publishing Limited, pp. 99–118.

Technology adoption during COVID-19. https://www.nielsen.com/uk/en/insights/article/2020/covid-19-the-unexpected-catalyst-for-tech-adoption/, Accessed on 28/05/2020.

The Guardian https://www.theguardian.com/sport/2020/feb/17/tokyo-marathon-restricted-to-elite-runners-over-coronavirus-scare-athletics, Accessed on 19/02/2020.

Torres, O., Swalhi, A., Mukerjee, J., Lasch, F. & Thurik, R. (2021). Health perception of French SME owners during the covid-19 pandemic, *International Review of Entrepreneurship*, 19(2), 151–168.

Wang, L. (2003). To compensate themselves: The consumption impulse in post SARS period. On WWW at http://www.nbjd.gov.cn, Accessed on 28/02/20.

WHO, Number of Novel coronavirus. https://www.who.int/emergencies/diseases/novel-coronavirus-2019, Accessed on 11/04/2022.

WHO. https://www.who.int/docs/default-source/coronaviruse/situation-reports/20200218-sitrep-29-covid-19.pdf?sfvrsn=6262de9e_2, Accessed on 19/03/2022.

Zhang, W., Gu, H. & Kavanaugh, R.R. (2005). The impacts of SARS on the consumer behaviour of Chinese domestic tourists. *Current Issues in Tourism*, 8(1), 22–38.

12

IT'S ALL ABOUT EXPERIENCES

The need for uniqueness in luxury consumption

Chen Ren, Brendan Paddison and Rebecca Biggins

Introduction

Luxury market sales are seeing a 5% annual growth reaching €49 billion in 2020, and the significant spending power has caused the marketing of luxury brands to focus upon fitting the expectations of consumers (Cho et al., 2022). Consumers have also developed the highest level of motivation for status consumption in luxury purchase, striving to improve their social standing through consumption. Luxury consumers are sensitive to peer reference and usually have a strong desire to convey a certain impression or conform to social norms (Kim and Jang, 2014). Such social standing or impression includes consumers' desire for a high level of uniqueness and distinction.

It is believed that consumption serves to improve consumers' social standing, especially luxury consumption, as luxury is closely associated with the idea of scarcity, uniqueness and exclusivity (Bhaduri and Stanforth, 2016). However, it seems that consumers' growing purchasing power is also creating a conflict with the level of the social distinction they strive to establish through luxury consumption. This emergent conflict is the result of luxury goods becoming much more affordable and accessible to new customers (Gardyn, 2002). Moreover, this demonstrates that there has been a shift in contemporary luxury consumption driven by the needs to improve social standing, with the new luxury no longer exclusive or rare enough to address consumers' needs for uniqueness (Kauppinen-Räisänen et al. 2018).

Despite previous studies which have attempted to understand the rise in luxury consumption through the lens of brand personality (Workman and Lee, 2011), brand consciousness (Grotts and Johnson, 2013; Gurău, 2012; Liao and Wang, 2009) and self-motivation perspectives (Truong, 2010; Mittal, 2006),

DOI: 10.4324/9781032637778-13

there is no clear answer or explanation of how luxury brands can satisfy the rising desire to obtain a high level of uniqueness (De Kerviler and Rodriguez, 2019; Giovannini et al., 2015) and that limited research has provided an in-depth understanding of consumers' attitudes and behaviours towards the notion of uniqueness consumption.

To address these gaps in the literature, this study explores consumers' growing need for uniqueness consumption through luxury purchases and draws on the Need for Uniqueness (NFU) framework to explore how luxury purchases are used to establish consumers' social distinction. The NFU framework suggests that there are three patterns usually followed by a consumer to achieve their needs for uniqueness consumption: (1) creative choice (purchases that are considered good choices by others), (2) unpopular choice (purchases which risks social disapproval) and (3) avoidance of similarity (avoid common products purchases) (Tian et al., 2001). Guided by this framework, this study critically explores how consumers satisfy their needs for uniqueness through luxury consumption, while also considering the extent to which consumers follow any of the three patterns to acquire a certain level of uniqueness.

This chapter therefore advances the existing theoretical understanding of luxury brand consumption by focusing on the idea of uniqueness luxury consumption. In order to contribute to existing knowledge on NFU in luxury consumption, this chapter presents findings from qualitative research which involved two focus group and 30 in-depth semi-structured interviews. Drawing on existing research, we further present and discuss the themes of luxury consumption patterns, satisfying NFU through luxury consumption and the approach of ethical luxury consumption in satisfying NFU. We also propose a new way of defining 'luxury' in the conclusion, capturing the shift in contemporary luxury consumption which has recently been driven by consumers: from material luxury consumption to experience luxury consumption.

Consumers' need for uniqueness

Consumers' NFU framework is grounded in Snyder and Fromkin's (1977) uniqueness theory, defined as the 'trait of pursuing differences relative to others through the acquisition, utilisation, and disposition of consumer goods for the purpose of developing and enhancing one's self-image and social image'. Therefore, the consequences of displaying uniqueness can be perceived as positive and such uniqueness is achieved not only through physical appearance, but also through symbolic meanings or publishing symbolic importance (Soh et al., 2017). People pursue a certain level of difference or uniqueness to acquire extrinsic rewards from the society or to obtain intrinsic satisfaction from themselves to be different (Snyder, 1992).

Consumers practise and satisfy their needs for uniqueness through consumption of goods and display such goods in order to enhance consumer's self-image and social image. The self-image enhancement process is highly personal, and it happens when the symbolic meaning of the goods is recognised by the consumers and is seen as part of the consumers' self-identity. Further, when the symbolic meaning of the purchased goods is recognised not only by the consumers themselves, but also publicly, especially if the symbolic meaning is appraised by others, it can be argued that consumers' social image is enhanced (Belk, 1988). Accordingly, Tian et al. (2001) conceptualised consumers' need for uniqueness into three dimensions: creative choice, unpopular choice and avoidance of similarity.

Creative choice: To be considered good choices by others

Displaying material goods is usually used to create consumers' self-image and identity. The uniqueness of the image and identity is reflected through the originality and uniqueness of the consumer goods (Belk, 1988). Especially when the choice of such consumer goods is considered good by others (social approval), for example when others think 'you are unique', the need for uniqueness is fulfilled. The choice to fulfil the needs of uniqueness via socially approved goods is a created choice. It reflects the idea of: *do they think I am unique enough?*

Unpopular choice: Take risks from social disapproval

In the event that consumers fail to pursue uniqueness through acquiring the socially approved uniqueness, the option to differentiate themselves is to challenge the 'social approval system' in the format of, for example breaking rules or challenging existing social norms, in order to enhance self-image, social identity and ultimately, satisfying their needs for uniqueness. Arguably, breaking rules or challenging existing social norms by themselves could satisfy the desire to 'stand out' (Tian et al., 2001; Ziller, 1964). It reflects the idea of: *people have poor taste and I am different!*

Avoidance of similarity: Move away from the norm

Uniqueness created through creative choice or unpopular choice has the risks of diminishing over time as the originally created uniqueness could become widely popular (Heckert, 1989). Therefore, the effectiveness of creative and unpopular choices is not sustainable in enhancing consumers' self-image and social identity. Consequently, the group of consumers who are pursuing a higher level of uniqueness are often monitoring others' consumed goods and avoiding displaying similar uniqueness, in order to recreate the differentness (Tian et al., 2001). It reflects the idea of: *I am keeping my eyes on you to be different to you!*

Satisfying NFU through luxury consumption

Predominantly luxury is associated with limited supply, high monetary value and superior quality (Bhaduri and Stanforth, 2016). Symbolic meanings of luxury or luxury consumption, drawing from the brands' heritage, craftsmanship, authenticity exclusive and expertise have also been highlighted, as consumers use this way of consumption as a display of status and wealth, and more importantly, as a way to reflect their personal and social identity (Tynan et al., 2010; Vigneron and Johnson, 2004). Owning a limited supplied item can be viewed as a power or status symbol and enhances individuals' self-image. Therefore, consumers do expect that a great level of uniqueness could be obtained through luxury consumption in order to enhance their personal and social identity (Liu, et al., 2019; Jin et al., 2013).

Possessing relatively scarce and valuable products is a socially accepted manner of expressing one's uniqueness and seems luxury consumption qualifies to satisfy consumers' desire for uniqueness (Lee, et al., 2021; Cheema and Kaikati, 2010; Amaldoss and Jain, 2005). Guided by the NFU framework, a luxury brand, including its products, versions or styles, can also satisfy consumers' desire of obtaining uniqueness through three ways: creative choice, unpopular choice and avoidance of similarity (Cho, et al., 2022; Chevalier and Mazzalovo, 2021; Tian et al., 2001).

Firstly, following the definition of creative choice, consumers' seeking for social differences can be achieved when their luxury purchase is considered good choices by others. As luxury brands usually reflect the idea of unique and unusual design, the desire for social distinction could be enhanced when the unique or unusual features of the luxury purchase are appreciated by others. The unique and unusual design sometimes comes with a certain level of risk (others may agree or disagree with your luxury taste). However, such risk is to accelerate consumers' evaluation of the brand uniqueness (Jebarajakirthy and Das, 2021; Kron, 1983; Snyder and Fromkin, 1977).

Secondly, following an unpopular choice in luxury consumption means that consumers seek for differences through selecting brands that risk social disapproval. In this case, a luxury brand may not necessarily challenge appropriate social manners but lack public popularity due to its unique design. Consumers actually perceive such 'unpopular' luxury brands as self-image or social image enhancing, as following these unpopular brands could exhibit a certain level of innovation and leadership (Jebarajakirthy and Das, 2021; Heckert, 1989).

Thirdly, avoiding similarity suggests that consumers who pursue a high need for uniqueness often monitor others' purchase to avoid similarity and to avoid the purchase of common products or brands (Tian et al., 2001). In the context of luxury consumption, a limited-edition luxury item or a discontinued luxury item is more attractive to establish one's uniqueness, as such initially unpopular items could gain acceptance over time (Jebarajakirthy and Das, 2021; Lee, et al., 2021; Heckert, 1989).

Methodology

This study was conducted by adopting a qualitative, interpretive approach based upon the premise that consumers have their own understanding of their situation or behaviour which they share within their responses (Veal, 1997). The data for the study was gathered via two focus group and 30 in-depth semi-structured interviews. Interviews are a common method for investigation in social research (Rubin and Rubin, 1995). However, it can be said that interviews can be restrictive. As Creswell (2012) cautions, interviews only allow the researcher to gather responses that have in some way been filtered by the judgement of the researcher through the building of the interview schedule, which is mainly influenced by the literature review. In order to address this issue, focus groups were conducted before the interviews to allow for a wider understanding of issues that may be relevant to include in the interview schedule.

Focus groups

In the first phase of the research, two focus groups of 12 and 15 participants were conducted. The number of participants required for each focus group was determined according to prior research conducted by Berns (2007) and Malhotra and Birks (2006). The participants were recruited using a convenience sampling method (Henry, 1990) using university research networks ('students as researcher scheme').

The two focus group interviews were conducted face-to-face in a pre-booked interview room on the premises of the University where the first author is a faculty member. The duration of each focus group interview was approximately two hours. A discussion guide was used to ensure effective flow of conversation and assist the moderator to conduct the focus group interview in a smooth manner (Greenbaum, 2000). The discussion guide covered two main themes which included (1) 'discussion on luxury consumption' and (2) 'discussion on NFU consumption pattern'. While the conversations were centred around the pre-developed talking points, new themes such as 'doing rather than buying', 'memorable luxury experiences' and 'uniqueness luxury experiences' were highlighted in the conversation. These new themes were used to guide the development of questions for the second phase of data collection that used the interview method.

The respondent profile of the focus group is presented in Table 12.1.

Interviews

A total of 30 interviews were conducted. The interview schedule was shaped by the result of the focus groups as well as the review of existing literature. As a result, the interview schedule comprised of themes including (1) the meaning of luxury, (2) luxury consumption pattern, (3) financial management,

TABLE 12.1 Sociodemographic characteristics of the sample of focus groups

		Percentage
Gender	Male	26%
	Female	74%
Age	18–24	55%
	25–34	25%
	35–44	10%
	45–54	10%
Education	Graduate degree (BSc, BSc Hons, etc.)	30%
	Postgraduate degree (MSc/MBA/PhD, etc.)	70%

(4) consumer profile, life stages, references groups and (5) current consumer culture. There were also additional questions included around the theme of 'needs for uniqueness', which explored participant's general view on uniqueness and their perceptions towards material and social uniqueness.

Each interview lasted approximately 60 minutes. To avoid the duplication of responses, the interview respondents were not the same as those in the focus groups. Interview participants were recruited as volunteers through University research networks.

The respondent profile of the interview is presented in Table 12.2.

Data analysis procedure

Data collected from the focus groups and interviews was analysed using the thematic analysis method. Thematic analysis attempts to identify and describe patterns and themes within the gathered data set (Braun and Clarke, 2021). All focus group and interview data were firstly transcribed in full verbatim in Microsoft Word and later transferred to the Nvivo8 software package for detailed analysis. Two independent coders then proceeded to analyse the data

TABLE 12.2 Sociodemographic characteristics of the sample of individual interview

		Percentage
Gender	Male	50%
	Female	50%
Age	18–24	37%
	25–34	38%
	35–44	10%
	45–54	15%
Education	Graduate degree (BSc, BSc Hons, etc.)	73.3%
	Postgraduate degree (MSc/MBA/PhD etc.)	26.7%

using a tiered coding system (Anderson, 2009). Firstly, Axil coding was carried out by searching for key themes. Secondly, selective coding was used where all key themes/categories identified were centred around central themes that help to achieve the research objectives and answer the research questions.

Findings and discussion: Ways to satisfying NFU through luxury consumption

The purpose of this chapter is to understand how luxury brands satisfy the rising desire to obtain a high level of uniqueness and to provide an in-depth understanding of consumers' attitudes and behaviours towards the notion of uniqueness consumption. Our findings revealed that consumers have developed a different luxury consumption pattern, supported by financial resource providers. The findings also showed that consumers have an increased desire to be unique and seek to display uniqueness through experienced-based luxury consumption and ethical luxury consumption, especially when material-based luxury goods are much more affordable and accessible (Gardyn, 2002). A third theme identified is that a new way of defining luxury featuring the ideas of 'experience, personal story and individual interpretation' is proposed to help to strengthen the social image and identify further. Figure 12.1 summarises the main findings of this research, which includes the overarching theme and descriptive and initial codes.

The affordable and approachable luxury

Increasingly, it is argued that consumers have more opportunities and reasons to spend (Fingerman et al., 2009). Consequently, consumers feel encouraged to engage with luxury consumption from a younger age. Several younger participants in this study, who have received financial support from their 'helicopter parents' (parents who hover and micromanage offspring's lives by being ultra-protective and supportive), agreed that luxury is much more approachable than before. For example, participants commented that luxury items are widely seen among them:

> Luxury is much more approachable for our generation.
>
> *(Participant 4)*

> I am not surprised that almost everyone from my class owns some luxury items.
>
> *(Focus Group 2)*

Further, participants in this study confirmed the role of parents providing financial support in their luxury consumption:

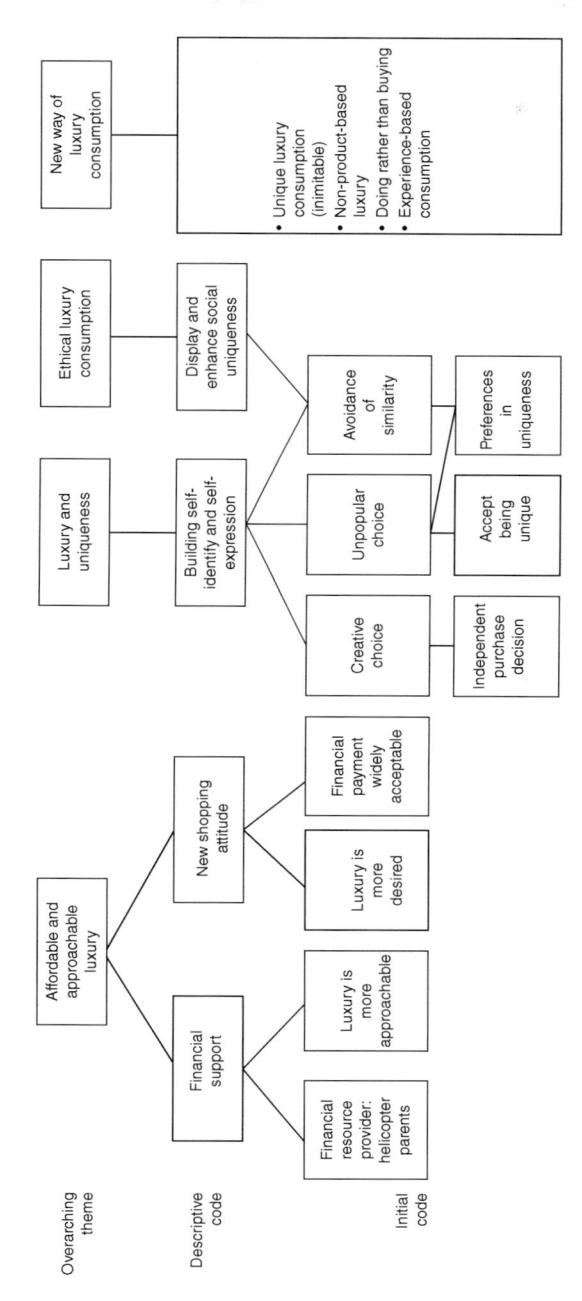

FIGURE 12.1   Summary of the main findings/themes of the research.

Well, my parents are going to pay for my trip to India after uni, I know I am spoiled, but I feel everyone's parents are paying for something, as a treat after our 3 years of hard work at uni.

(Focus Group 1)

A lot of my friends are all driving nice (luxury) cars, mum and dad paid as gifts. I know a few of them who got BMW when they finished university.

(Participant 10)

It seems that in this case some consumers, especially consumers from younger generations, are not matching their spending with their financial affordability (Kim and Jang, 2014). Instead, they have developed a certain attitude that they are entitled to accept financial help from their parent to enjoy luxury and luxury purchases are therefore becoming more readily available. Consequently, younger consumers are portrayed as impatient and as having an 'I want it all and I want it now' attitude (Bolton et al., 2013; Barton et al., 2013; Yarrow & O'Donnell, 2010).

Such received financial support has encouraged younger consumers to take on certain levels of financial risks in luxury consumption. For example, to obtain certain luxury items having debt or short loans is widely acceptable for some participants in this study:

Yes my BMW is on finance.

(Participant 6)

Luxury is to have a newly fitted kitchen…Usually these kitchen companies are a bit more expensive, but it's fine, I will go for finance, it's zero interest and only cost me around £300 per month.

(Participant 1)

The received financial support has also given younger consumers a sense of security to take on debt or loans: '*I knew a few people in my age whose mortgage is looked after by their parents*' (Participant 7). It seems that consumers could develop their attitude towards money as that wealth and material goods are to be desired (Butcher et al. 2017; Lissitsa and Kol, 2016; Giovannini et al., 2015; Bakewell and Mitchell, 2003), which encouraged them to pursue luxury consumptions in a more aggressive way:

Oh come on, everyone in our age has some luxury items!

(Focus Group 2)

mum and dad will help me out.

(Participant 20)

Luxury and uniqueness

Consumers are found to build self-expression and self-identification through associating themselves with the idea of 'being unique' and that the consequences of being unique appear to be highly positive. Consumers welcome individuals with a certain level of uniqueness and they would rather accept someone with a *'unique personality and a bit different'* (Participant 17) to someone with *'no personality at all'* (Participant 15). Hence, it seems that being unique and distinct from others is a way to draw self-confidence for consumers:

> they (people who have the desire to be distinctive) look cool.
>
> *(Participant 25)*

> I feel people without a personality can't blend in, they are dull. We'd like a more diverse group of people around us.
>
> *(Participant 15)*

The display of uniqueness, in general, is not only through individuals' physical appearance but also symbolic meanings or publish symbolic importance (Soh et al., 2017). It was found that participants in this study prefer social norms uniqueness over material uniqueness. They suggested that *'the real unique thing is not what you can see, but something in your mind'* (Participant 25). Although most people prefer material uniqueness as a less risky way to display their status, it appears that others are not scared of taking on more risks to display their social norms uniqueness: *'being unique is to think and behave differently, to express the unique yourself, rather than using an expensive handbag to standout'* (Participant 13). This finding echoes Snyder and Fromkin's (1977) theory of uniqueness, which defines unique as being distinct from others, especially when the individual is perceived as having a higher level of uniqueness in terms of social status.

As discussed in the previous section, possessing luxury products has been recognised socially as a manner to satisfy individual's need for uniqueness (Amaldoss and Jain, 2005; Tiam et al., 2001). It seems that some have developed a different way of satisfying their own NFU, not through possessing or displaying luxury items, but through experiencing and experimenting:

> Yes they can buy a luxury item and they can share a luxury handbag on social media, but not what I experienced and learnt from the gap year trip to the States. That experience is the real luxury.
>
> *(Focus Group 2)*

Specifically, the word luxury is viewed as a commercial phrase used to push consumers to purchase: *'people pay too much for the logo, they are paying too much for the word of "luxury"'* (Participant 22). Some even see such physical attributes of the luxury brand (i.e., price and logo) decrease the value of brand:

'*Well, I feel a luxury item is not that luxury anymore with any logo on it*' (Participant 22). It seems that the level of uniqueness gained through acquiring a luxury item is regarded as very low and cannot satisfy consumers' NFU anymore. Instead, it is found that participants think 'luxury' should be an expression used to communicate something significant or unique about the good or service (Danziger, 2013). For example, when searching for uniqueness or distinction, they favour the types of luxury that show a great level of symbolic value, including experiences or adventures that bring them closer to the life they want to live. They want the luxury purchase to satisfy their needs fully so they can legitimise the expense. These experiences or adventure-based luxury consumptions could be: '*a spa trip with your besties*' (Participant 3), '*my 3-year of uni life…It's unique, it's my own 3 years of experience. Nobody can copy it*' (Participant 7). Sometimes the luxury consumption comes with little expenses: '*luxury is 2 tickets for me and my Dad to watch footy on a Saturday afternoon. That experience has no logo on it, but it is luxury and as unique as you can imagine*' (Participant 17).

In the previous section, we argued that a luxury brand can satisfy consumers' NFU through three ways of creative choice (when luxury purchase is considered good choices by others), following an unpopular choice (choosing luxury brand which risks social disapproval) and avoiding similarity (avoid purchasing of common brands). Markow (2005) suggests that some worry more about how other people think of them. However, our findings suggest the opposite. It was found that 'creative choice' has not been followed by some as they see luxury consumption as a pure independent decision and that influence from others is less: '*luxury is a quite individual thing for us, why should we worry about other people's opinions on something that is very unique to me*' (Participant 19). Further, participants are not afraid to follow an unpopular choice, instead, they are trying to maximise the level of uniqueness through luxury consumption through avoiding similarity: '*(In luxury consumption) if I can afford to be unique, I will be unique. Being unique is cool. We don't like to be the follower*' (Participant 30). Clearly an 'unpopular' luxury item/brand has been labelled 'cool' and seen as a self-image or social image-enhancing tool. This finding contrasts with the previous findings in this area which show that consumers follow the purchases of their peers to keep up with trends and avoid making incorrect decisions (Fernandez, 2009, Twenge and Campbell, 2008). It seems that the idea of 'use luxury consumption to express uniqueness' has been pushed further, and people are separating their luxury consumption decisions with the influence of any reference groups, to obtain a higher level of uniqueness.

The new way of luxury consumption

Not only avoiding purchasing popular luxury item/brand is less powerful to achieve the level of uniqueness consumers desires, it was found that purchasing a limited edition luxury item or a discontinued luxury item can fail to satisfy consumers' NFU luxury consumption:

> No luxury item is that unique any more, even the limited edition ones, there will be someone who bought the same product.
>
> *(Focus Group 2)*

Therefore, while materials uniqueness cannot satisfy such rising demands, consumers are looking for new ways to define themselves through a certain type of uniqueness that others cannot easily copy:

> maybe something else, such as a luxury experience, a feeling of luxury, can give me a sense of uniqueness.
>
> *(Focus Group 1)*

Our findings suggest that the traditional product-based luxury items are not able to provide the level of 'uniqueness' and a new way of defining luxury is proposed by participants. This new way of defining luxury consumption is less materialistic focused and luxury items are classed as dispensable:

> consumers are looking for something beyond the product itself, so they (luxury shops) give you a drink in the shop, they try to have a conversation with you, they try to create a nice experience for you. They want you to remember this experience.
>
> *(Focus Group 1)*

These new ways of luxury consumption reflect the idea of 'doing' rather than 'owning' which suggests that consumers want to personally involve in creating the luxuriousness, in order to build their personal experience and story for sharing and memorising. It was explained that:

> Luxury to me, is being able to DO whatever I want to do.
>
> *(Participant 4)*

> luxury is not an item, but something you can always share and talk about.
>
> *(Participant 18)*

It seems that consumers are valuing luxury differently, from a purely transactional relationship to a holistic experience creation (Atwal and Williams, 2009).

In summary, while previously consumers have the passion to maintain an iconoclastic worldview and transformed the luxury market from a conspicuous consumption model to a more individualistic type, it seems that such transformation has been advanced by defining today's luxury as a way to 'celebrate of personal creativity, expressiveness, intelligence, fluidity and above all, meaning' (Dumoulin, 2007). Fundamentally, the newly developed luxury category is experience based and it highlights the idea of 'doing' rather than 'buying' to gain a high level of uniqueness through consumption. A 'doing'-based personal experience cannot be copied or compared, and the level of individualism gained is high enough to allow consumers to create their own unique stories.

Ethical luxury consumption and NFU

Material goods are used by many consumers as a way of representing their individuality and significance, and furthermore, they are used as tools to contribute to the construction of self and self-identity (Wattanasuwan, 2005; Belk, 1985). Similarly, consumers could use consumption of goods and services to display their personal values, characteristics and identity (Kjeldgaard and Askegaard, 2006). For example, consumption of ethical luxury brands is used as a way to build consumers' identity of being 'responsible' and sensitive to sustainability:

> This is definitely a 'thing' for us nowadays, we want to be identified as the generation who takes on responsibilities and I'd like to be part of it.
>
> *(Participant 7)*

However, while previous research suggests that consumers nowadays display a high level of sensitivity to sustainability than any other generations and that consumers can even hate brand who lack ethics and integrity (Marticotte, 2018), Kapferer and Michaut-Denizeau's (2020), more recent studies suggest quite differently that '*sustainability is not at the front of their minds when buying luxury' (p.45)*. The reason might be that there is a clear difference between the value a 'luxury brand' is presenting and the value a 'sustainable luxury brand' is hosting. For example, the nature of luxury brands being a signal of status and uniqueness has attracted consumers who prioritise their self-enhancement values, while initiatives of building a sustainable luxury brand prioritises protecting others' welfare and considering the interests of others (Park et al., 2008). Participants in this study admitted that they evaluate the benefit (self-enhancement) gains from luxury consumption prior to evaluate if such luxury brand is ethical:

> I do prefer an ethical luxury brand, but it has to be a luxury brand first and worth the money I pay for.
>
> *(Participant 7)*

Furthermore, it seems that ethical luxury consumption decision-making is not made to protect others' welfare, and to consider the interests of others, consumers think associating with ethical luxury brands help them to stand out more and to satisfy their NFU:

> buying from ethical brands is trendy now, as if everybody is saying: look at me, look at me, I buy from ethical brands, I am different! But do we really check the authenticity of these brand and check if these brands are doing what they promised? Maybe not.
>
> *(Participant 11)*

Conclusion

Having responded to the call for research to examine how luxury brands satisfy the rising desire to obtain a high level of uniqueness (De Kerviler and Rodriguez, 2019; Giovannini et al., 2015) and to provide an in-depth understanding of consumers' attitudes and behaviours towards the notion of uniqueness consumption, the main finding of this research suggests that first of all, luxury consumption is needed and used as a form of self-expression, although such strive to establish their social distinction through luxury consumption has been challenged by the fact that luxury goods are much more affordable and accessible (Gardyn, 2002).

More specifically, firstly, guided by the NFU framework, it was found that consumers have developed a new luxury shopping pattern to satisfy their desire to obtain a greater level of uniqueness and social distinction. They prefer a symbolic and social norm-based uniqueness, developed through a new way of experienced or adventure-based luxury consumptions (i.e., a trip to a foreign country rather than an expensive handbag that others could obtain). When making such an experience or adventure-based luxury consumption decision, consumers tend to consider others' influence less and claim that such consumption should be an individual decision. This contradicts existing research which suggest that consumers could follow their peers to keep up with trends and avoid making wrong decisions (Fernandez, 2009; Twenge and Campbell, 2008). Therefore, consumers these days are declining to satisfy their NFU through creative choice (when luxury purchase is considered good choices by others), but preferring following an unpopular choice (choosing luxury brand which risk social disapproval) and avoiding similarity (avoid purchasing of common brands) to maximise the level of uniqueness.

Secondly, we found that consumers do consider choosing ethical luxury brands in consumption but '*sustainability is not at the front of their minds when buying luxury*' (Kapferer and Michaut-Denizeau, 2020, p.45). A dominating reason of ethical luxury consumption could be that people associate with ethical luxury brands to gain social distinction to satisfy their NFU further.

Finally, we found that there has been a development of a new way of defining luxury consumption, which could be summarised as:

> luxury consumption is a less materialistic focused, but more experience focused way of consumption which usually involves a personal story or an individual interpretation. The personal meaning associated with luxury consumption is to satisfy the increasing needs of obtaining uniqueness to build consumers' image in the society.

References

Amaldoss, W. and Jain, S. (2005) Conspicuous consumption and sophisticated thinking. *Management Science*, 51 (10), pp.1449–1466.

Anderson, V. (2009) *Research Methods in Human Resource Management*. Chartered Institute of Personnel and Development, London.

Atwal, G. and Williams, A. (2009) Luxury brand marketing – The experience is everything. *Journal of Brand Management*, 16 (5–6), pp.338–346.

Bakewell, C. and Mitchell, V. (2003) Generation Y female consumer decision-making styles. *International Journal of Retail & Distribution Management*, 31 (2), pp. 95–106.

Barton, D., Chen, Y. and Jin, A. (2013) *Mapping china's middle class*. McKinsey.

Belk, R.W. (1988) 'Possessions and the Extended Self,' *Journal of Consumer Research*, 15 (9), pp.139–168.

Belk, R.W. (1985) Materialism: Trait aspects of living in the material world. *Journal of Consumer Research*, 12 (3), pp.265–280.

Bhaduri, G. and Stanforth, N. (2016) Evaluation of absolute luxury. *Journal of Fashion Marketing and Management: An International Journal*, 20 (4), pp.471–486.

Bolton, R.N., Parasuraman, A., Hoefnagels, A., Migchels, N., Kabadayi, S., Gruber, T., Komarova Loureiro, Y. and Solnet, D. (2013) Understanding generation Y and their use of social media: A review and research agenda. *Journal of Service Management*, 24 (3), pp.245–267.

Braun, V., & Clarke, V. (2021) *Thematic Analysis: A Practical Guide*. SAGE Publications.

Buckle, C. (2019) *The luxury market in 2019: What brands should know* [Internet]. Available from https://blog.globalwebindex.com/chart-of-the-week/luxury-market-2019/.

Butcher, L., Phau, I. and Shimul, A.S. (2017) Uniqueness and status consumption in generation Y consumers. *Marketing Intelligence & Planning*, 35 (5), pp.673–687.

Cheema, A. and Kaikati, A.M. (2010) The effect of need for uniqueness on word of mouth. *Journal of Marketing Research*, 47 (3), pp.553–563.

Cho, E., Kim-Vick, J., & Yu, U.-J. (2022) Unveiling motivation for luxury fashion purchase among Gen Z consumers: need for uniqueness versus bandwagon effect. *International Journal of Fashion Design, Technology and Education*, 15 (1), pp.24–34.

Creswell, J.W. (2012) *Qualitative inquiry & research design: Choosing among five approaches *. 3 ed. London, SAGE Publications.

Danziger, K. (2013) Psychology and its history. *Theory & Psychology*, 23 (6), pp.829–839.

de Kerviler, G. and Rodriguez, C.M. (2019) Luxury brand experiences and relationship quality for millennials: The role of self-expansion. *Journal of Business Research*, 102, pp.250–262.

Debevec, K., Schewe, C.D., Madden, T.J. and Diamond, W.D. (2013) Are today's millennials splintering into a new generational cohort? maybe. *Journal of Consumer Behaviour*, 12 (1), pp.20–31.

Dotson, M.J. and Hyatt, E.M. (2005) Major influence factors in children's consumer socialization. *Journal of Consumer Marketing*, 22 (1), pp.35–42.

Dries, N., Pepermans, R. and De Kerpel, E. (2008) Exploring four generations' beliefs about career. *Journal of Managerial Psychology*, 23 (8), pp.907–928.

Dumoulin, D. (2007) What is today's definition of luxury? *Admap*, 3 (481), pp.27–29.

Fernandez, P.R. (2009) Impact of branding on gen y's choice of clothing. *Journal of the South East Asia Research Centre for Communications and Humanities*, 1 (1), pp.79–95.

Festinger, L. and Carlsmith, J.M. (1959) Cognitive consequences of forced compliance. *The Journal of Abnormal and Social Psychology*, 58 (2), pp.203–210.

Fingerman, K., Miller, L., Birditt, K. and Zarit, S. (2009) Giving to the good and the needy: Parental support of grown children. *Journal of Marriage and Family*, 71 (5), pp.1220–1233.

Gardyn, R. (2002) *Oh, the good life*. Trans.Anonymous. Detroit, Crain Communications, Incorporated.

Giovannini, S., Xu, Y. and Thomas, J. (2015) Luxury fashion consumption and generation Y consumers. *Journal of Fashion Marketing and Management: An International Journal*, 19 (1), pp.22–40.

Greenbaum, T.L. (2000) *Moderating Focus Group: A Practical Guide for Group Facilitation*. Sage Publications, London.

Grotts, A.S. and Johnson, T.W. (2013) Millennial consumers' status consumption of handbags. *Journal of Fashion Marketing and Management*, 17 (3), pp.280–293.

Gurău, C. (2012) A life-stage analysis of consumer loyalty profile: Comparing generation X and millennial consumers. *Journal of Consumer Marketing*, 29 (2), pp.103–113.

Gustafson, K. (2015) *Millennials redefine luxury—and the stakes are high* [Internet]. Available from https://www.cnbc.com/2015/02/18/-redefine-luxury-and-the-stakes-are-high.html.

Hammersley, M. and Atkinson, P. (1995) *Ethnography*. 2nd ed. ed. London u.a, Routledge.

Heckert, D.M. (1989) The relativity of positive deviance: The case of the french impressionists. *Deviant Behavior*, 10 (2), pp.131–144.

Henry, G.T. (1990) *Applied social research methods: Practical sampling*. Thousand Oaks, CA, SAGE Publications.

Holstein, J.A. and Gubrium, J.F. (1995) *The active interview*. Thousand Oaks [u.a.], Sage Publ.

Hunt, J. (2008) "Make Room For Daddy…And Mommy: Helicopter Parents Are Here!", *The Journal of Academic Administration in Higher Education*, Vol. 4, No. 1, pp.9–11.

Inglehart, R. (1997) *Modernization and postmodernization*. Princeton, N.J, Princeton Univ. Press.

Jay, E. (2012) *New breed of consumer shakes up luxury fashion* [Internet]. Available from https://www.mobilemarketer.com/ex/mobilemarketer/cms/opinion/columns/12361.html.

Jebarajakirthy, C. and Das, M. (2021) Uniqueness and luxury: A moderated mediation approach. *Journal of Retailing and Consumer Services*, Vol. 60, pp.102477.

Jin, L., He, Y., Zou, D. and Xu, Q. (2013) How affirmational versus negational identification frames influence uniqueness-seeking behavior. *Psychology & Marketing*, 30 (10), pp.891–902.

Kapferer, J. (1997) *Strategic brand management*. 2ed ed. London, Kogan Page.

Kapfere, J. and Michaut-Denizeau, A. (2020) Are millennials really more sensitive to sustainable luxury? A cross-generational international comparison of sustainability consciousness when buying luxury. *Journal of Brand Management*, 27, pp.35–47.

Kim, D. and Jang, S. (2014) Motivational drivers for status consumption: A study of generation Y consumers. *International Journal of Hospitality Management*, 38, pp.39–47.

Kjeldgaard, D. and Askegaard, S. (2006) The glocalization of youth culture: The global youth segment as structures of common difference. *Journal of Consumer Research*, 33 (2), pp.231–247.

Kron, J. (1983) *Home-psych: The social psychology of home and decorations*. New York, Potter.

Kupperschmidt, B. (2000) Tips to help you recruit, manage, and keep generation X employees. *Nursing Management (Springhouse)*, 31 (3), pp.58–60.

Lancaster, L.C. and Stillman, D. (2003) *When generations collide*. 1 ed. New York, NY, HarperBusiness.

Lee, M., Bae, J., & Koo, D.-M. (2021) The effect of materialism on conspicuous vs inconspicuous luxury consumption: focused on need for uniqueness, self-monitoring and self-construal. *Asia Pacific Journal of Marketing and Logistics*, 33 (3), pp.869–887.

Leone, D. (2015) *Luxury industry facing the millennials opportunity* [Internet]. Available from https://www.linkedin.com/pulse/luxury-industry-facing-the-millennials-opportunity-daniela-leone/ [Accessed August 1, 2015].

Liao, J. and Wang, L. (2009) Face as a mediator of the relationship between material value and brand consciousness. *Psychology and Marketing*, 26 (11), pp.987–1001.

Lissitsa, S. and Kol, O. (2016) Generation X vs. generation Y – A decade of online shopping. *Journal of Retailing and Consumer Services*, 31, pp.304–312.

Liu, H., Wu, L. and Li, X. (2019) Social media envy: How experience sharing on social networking sites drives millennials' aspirational tourism consumption. *Journal of Travel Research*, 58 (3), pp.355–369.

Markow, D. (2005) Children's reactions to tragedy. *Young Consumers*, 6 (2), pp.8–10.

Marticotte, F. (2018) *Why millennials hate some luxury brands*. In Research paper presented at the LVMH-SMU Luxury Conference, Singapore, May 11–12.

Meriac, J.P., Woehr, D.J. and Banister, C. (2010) Generational differences in work ethic: An examination of measurement equivalence across three cohorts. *Journal of Business and Psychology*, 25 (2), pp.315–324.

Mittal, B. (2006) I, me, and mine—how products become consumers' extended selves. *Journal of Consumer Behaviour*, 5 (6), pp.550–562.

Naumovska, L. (2017) Marketing Communication Strategies for Generation Y – Millennials. *Business Management and Strategy*, 8 (1), pp.123–133.

Okonkwo, U. (2007) *Luxury fashion branding*. Basingstoke, Hampshire [u.a.], Palgrave Macmillan.

Orlikowski, W.J. and Baroudi, J.J. (1991) Studying information technology in organizations: Research approaches and assumptions. *Information Systems Research*, 2 (1), pp.1–28.

Pam, D. (2015) *Five luxe trends for 2015* [Internet]. Available from https://unitymarketingonline.com/wp-content/uploads/Five-Key-Luxury-Market-Trends-FINAL-NEW.pdf.

Park, H., Rabolt, N.J. and Sook Jeon, K. (2008) Purchasing global luxury brands among young korean consumers. *Journal of Fashion Marketing and Management: An International Journal*, 12 (2), pp.244–259.

Peluchette, J.V.E., Kovanic, N. and Partridge, D. (2013) "Helicopter parents hovering in the workplace", *Business horizons*, Vol. 56, No. 5, pp.601–609.

Phau, I. and Prendergast, G. (2000) Consuming luxury brands: The relevance of the 'Rarity principle'. *Journal of Brand Management*, 8 (2), pp.122–138.

Purwanto, E., Deviny, J. and Mutahar, A.M. (2020) "The Mediating Role of Trust in the Relationship Between Corporate Image, Security, Word of Mouth and Loyalty in M-Banking Using among the Millennial Generation in Indonesia", *Management & Marketing*, vol. 15, no. 2, pp.255–274.

Rosen, M. (1991) Coming to terms with the field: Understanding and doing organizational ethnography. *Journal of Management Studies*, 28 (1), pp.1–24.

Rubin, H.J. and Rubin, I.S. (1995) *Qualitative interviewing: The art of hearing data*. 2nd ed. London, Sage Publication.

Saunders, M.N.K., Thornhill, A. and Lewis, P. (2019) *Research methods for business students*. 8th ed. Harlow, United Kingdom, Pearson Education.

Schewe, C.D. and Meredith, G. (2004) Segmenting global markets by generational cohorts: Determining motivations by age. *Journal of Consumer Behaviour*, 4 (1), pp.51–63.

Silverman, D. (2011) *Interpreting qualitative data*. 4 ed. London, Sage.

Silverstein, M.J., Fiske, N. and Butman, J. (2008) *Trading up*. Paperback ed. with a new introduction ed. New York, NY, Portfolio.

Snyder, C.R. (1992) Product scarcity by need for uniqueness interaction: A consumer catch-22 carousel? *Basic and Applied Social Psychology*, 13 (1), pp.9–24.

Snyder, C.R. and Fromkin, H.L. (1977) Abnormality as a positive characteristic: The development and validation of a scale measuring need for uniqueness. *Journal of Abnormal Psychology*, 86 (5), pp.518–527.

Soh, C.Q.Y., Rezaei, S. and Gu, M. (2017) A structural model of the antecedents and consequences of generation Y luxury fashion goods purchase decisions. *Young Consumers*, 18 (2), pp.180–204.

Strauss, W. and Howe, N. (1991) *Generations: : The history of america's future*. New York, Quill, Morrow.

Tian, K.T., Bearden, W.O. and Hunter, G.L. (2001) Consumers' need for uniqueness: Scale development and validation. *Journal of Consumer Research*, 28 (1), pp.50–66.

Tilford, C. (2018) *The millennial moment — in charts* [Internet]. Available from https://www.ft.com/content/f81ac17a-68ae-11e8-b6eb-4acfcfb08c11

Truong, Y. (2010) Personal aspirations and the consumption of luxury goods. *International Journal of Market Research*, 52 (5), pp.655–673.

Tuškej, U., Golob, U. and Podnar, K. (2013) "The role of consumer–brand identification in building brand relationships", *Journal of Business Research*, Vol. 66, No. 1, pp.53–59.

Twenge, J.M. (2014) *Generation me*. 2nd ed. New York, NY, Atria Publishers.

Twenge, J.M. and Campbell, S.M. (2008) Generational differences in psychological traits and their impact on the workplace. *Journal of Managerial Psychology*, 23 (8), pp.862–877.

Tynan, C., McKechnie, S. and Chhuon, C. (2010) Co-creating value for luxury brands. *Journal of Business Research*, 63 (11), pp.1156–1163.

Veal, A.J. (1997) *Research methods for leisure and tourism*. 2. ed. ed. London, Financial Times/ Prentice Hall.

Vigneron, F. and Johnson, L.W. (2004) Measuring perceptions of brand luxury. *Journal of Brand Management*, 11 (6), pp.484–506.

Wattanasuwan, K. (2005) The self and symbolic consumption. *Journal of American Academy of Business*, 6 (1), pp.179–184.

Wiedmann, K., Hennigs, N. and Siebels, A. (2009) Value-based segmentation of luxury consumption behavior. *Psychology and Marketing*, 26 (7), pp.625–651.

Workman, J.E. and Lee, S. (2011) Materialism, fashion consumers and gender: A cross-cultural study. *International Journal of Consumer Studies*, 35 (1), pp.50–57.

Yarrow, K. and O'Donnell, J. (2010) Gen Buy: How tweens, teens, and twenty-somethings are revolutionizing retail. *Journal of Consumer Marketing*, 27 (6), pp.564–565.

Young, A.M. and Hinesly, M.D. (2012) Identifying millennials' key influencers from early childhood. *The Journal of Consumer Marketing*, 29 (2), pp.146–155.

Ziller, R.C. (1964) 'Individuation and Socialization: A Theory of Assimilation in Large Organizations,' *Human Relations*, 17 (4), pp.341–360.

13

THE EFFECT OF SOCIAL MEDIA INFLUENCERS ON CONSUMER FOOD AND BEVERAGE CONSUMPTION BEHAVIOUR

Ece İpekoğlu and İrem Enser

Introduction

The rapid spread of social networks and the production of content by users on social media have brought new areas to the field of marketing and advertising. Advertising/promotion practices made by famous people through TV, radio or newspapers, which also exist in the traditional understanding of advertising, have been replaced by product promotions, evaluations and recommendations made by both celebrities and people known only on social media channels through social media (Geyser, 2023).

Instead of personal information, which is considered to be the most effective when compiling information in the consumer's decision process, this time, resources have turned into a new structure through people who exist on digital platforms and whose reliability is thought to have increased in parallel with the number of followers they have reached. This new structure is called 'influence marketing' (Brown and Fiorella, 2013). Consumers' desire to obtain information before purchasing a product and to benefit from the experiences of people whose opinions they trust is as valid today as it was in the past, and this process of obtaining information and sharing experiences has become much easier and faster thanks to some people who are accepted as competent in certain subjects on social media. It is observed that these influencers, who will be accepted as opinion leaders in their fields, have become an important communication element supported by new-era marketers. Thus, as a version of viral marketing, influence marketing studies, which are carried out by presenting the product experiences of well-known people with a high number of followers on social media in a natural environment, have recently become quite common.

DOI: 10.4324/9781032637778-14

In this chapter, the influences in food and beverage culture are investigated, which have been affected by social media and consumption culture, since consumer behaviour on traditional cuisine is affected by social media sharing of food and beverage influencers.

Influencer marketing: An effective marketing strategy

Traditional marketing developed for the effective and profitable delivery of what is produced to the customer techniques had to undergo change with the changes in communication and therefore in the world. Especially with the invention of the internet and the emergence of social media, 'influencer marketing' or 'phenomenon marketing' has become the ultimate point of this change (Brown and Hayes, 2008).

Influencer marketing, which is thought to be a new concept nowadays, in which influencers are also involved, in fact, it is not a new concept, it is a practice that has been done in different ways and with different tools.

In the context of influencer marketing nowadays, companies, celebrities, bloggers, vloggers, social media influencers (celebrities on Instagram and Twitter, for example), and others are followed on different social media platforms. The quantity of followers engages in a variety of cooperative activities with others.

Social networks, which are indispensable for modern Influencer Marketing, are becoming more and more popular with the new generation of consumers. It gains meaning and is becoming increasingly important. In this direction, individuals can reach any product or service through their social networks (Brown and Hayes, 2008).

Before making a purchasing decision, they shape their purchasing decision by considering the opinions and suggestions of other users in these social networks. These people not only perform the purchasing behaviour but also influence the purchasing decision of other users who are interested in that product or service and who are considering purchasing it. For this reason, many brands have started to benefit from influencer marketing in order to create customer loyalty and trust.

Brown and Hayes (2008: 50) define the concept of influencer as a third party that shapes consumers' purchasing decisions in a meaningful way. They express themselves as individuals. Influencers (phenomena) within social media platforms, celebrities, internet phenomena, bloggers, YouTube vloggers, Instagram celebrities, etc. have the ability to persuade people who have a very high potential to influence the people around them. Any brand is conditioned to make a recommendation for a product or service, these individuals are then able to make a recommendation to a person who is loyal to them. They have an audience of followers and these audiences trust and believe in their posts and benefit from their experiences.

In today's digital age, social networks that are rapidly changing and developing day by day, Influencer Marketing has been effective in bringing the concept to the forefront. By influencing the masses in the digital age, a very large Influencer Marketing, which has caught a breakthrough, has come to the fore as a rising marketing technique of the digital age has succeeded.

Brown and Fiorella (2013) explain the mentality of influencer marketing as Fisherman's Influence Model. The main objective of this model is to 'spread the widest net to catch the most fish'. The implementation of this favours the influencers with the largest number of followers and creates the greatest brand awareness and eventually is the realisation of purchasing behaviour. Influencers can share the brand's messages or suggestions widely and allow large masses to share it in their own social environment. As can be seen, marketing managers need to be able to represent themselves and to be able to communicate their products or services to a wider audience. They choose influencers who can introduce them to the masses (Brown and Fiorella, 2013: 77–78).

In the study by Ye, Hudders, De Jans and De Veirman (2021), they conducted a bibliometric examination of scholarly literature concerning influencer marketing in order to create an overview of the field. Their investigation led them to identify 387 articles that centre on influencer marketing, with the earliest publication dating back to 2003. Among these articles, the *International Journal of Advertising* was ranked as the sixth most prolific journal in terms of contributions to influencer marketing research, having published seven articles that had accumulated a total of 74 citations at the time of the study's publication (Ye et al., 2021). An analysis of these publications uncovered five recurring themes in influencer marketing research, which encompassed: (1) the effectiveness of influencer marketing, (2) viewpoints of various stakeholders on influencer marketing, (3) the application of influencer marketing in specific product categories and industries, (4) the processes of identifying, selecting and engaging influencers and (5) considerations of ethical concerns and disclosure practices within the realm of influencer marketing. In the following section, social media influencers are presented as marketing figures.

Social media influencers

An influencer is a person who has a large number of followers on social media and whose products from brands speak for the brand or the brand's products on their own social media accounts a person who receives a certain fee, free products, holiday gift in order to post them (Kadekova and Holiencinova, 2018). The social networks where influencers are most popular and most used are Facebook, Instagram, Snapchat and YouTube. Influencers use their position, authority, industry connections, and knowledge to assist firms they endorse on social media, which can have an impact on consumers' purchase decisions. These people are not only a marketing tool but rather they are valuable as they enable brands to build social relationships with customers.

Influencers are almost recognised as the phenomenon of the late modern era. In recent years, the number of marketing agreements made by brands with these people has increased considerably (Olenski 2017). In this sense, the groups that brands use to influence consumers for advertising are grouped under four categories: celebrities, industry experts and opinion leaders, bloggers and content producers, and influencers. Industry experts and opinion leaders are important for some brands depending on the sector. Famous people can already be considered as natural influencers and although their preference rates are decreasing compared to the past, they still maintain their effectiveness in this field. Bloggers, content producers and influencers are the groups that establish the most authentic and active relationships with their followers on social media today. Brands are also aware of this and support this communication (Geyser, 2023).

Social media influencers (SMIs) typically establish themselves as experts in a specific niche. For instance, consider Arif Doğan from Turkey, an influencer renowned for his expertise in food. With a staggering 1.7 million Instagram followers, he has achieved immense success. Notably, Arif Doğan's journey to becoming a highly regarded influencer can be attributed to his personal and traditional recipe advice from Turkish cuisine that he shares with his millions of followers. Similarly, Refika Birgül, a famous Turkish chef, boasts over 1.5 million Instagram followers. She has successfully cultivated her personal recipes and collaborated with prestigious companies like Knorr and İçim through sponsorships.

Food and beverage culture and behaviour of consumers

Before the digital era, food and beverage culture was mostly conducted by traditional recipe books. In line with the emergence of the internet, these recipe books took place as 'food blogs' and bloggers' effect on food and beverage culture became prominent. The behaviour of food and beverage consumers is changing according to changing consumption habits, rapidly developing technology, globalisation and consumers' search for innovation. There have been important and radical changes in social, economic, technological and cultural aspects from the 16th century to the 21st century. The change experienced during this period also affected the traditional food and beverage culture (Aksoy and Üner, 2016). In the meantime, the consumers' food and beverage consumption behaviour shifted culturally, as Riley (1994) mentioned. Riley (1994) argues the food and beverage market is a holistic experience rather than a simple market segmentation. The food and beverage industry, which can be considered as a short-lived fashion product from a postmodern perspective (Johns and Pine, 2002), has begun to be considered as a complete experience rather than a service.

New gastronomic trends become more visible in food and beverage culture, due to fast-changing consumption habits. In some cases, these new trends may overshadow traditional food and beverage culture (Türker and Süzer, 2022).

The constant change in people's eating habits or the increase in the expectations of individuals from the food and beverage sector brings along new trends. From the earliest times of human existence to the present day, eating and drinking habits have constantly changed, and this has led to the emergence of different culinary trends in certain periods. Particularly in the 17th century, instead of merely meeting essential needs, the tendency to eat, with the desire to taste different pleasures and experiences, has enabled the emergence of many different culinary trends until today (Erdem and Akyürek, 2017). Social media has a significant impact on shaping the reproduction of food and beverage culture. The marketing strategies planned by food and beverage businesses in social media are encouraging people to consume more even if they have no plan to do. The social media has an important impact on society's consumption habits (Acar, Çizmeci and Turan, 2021). Especially in recent years, problems have arisen as a result of people gradually moving away from natural and local nutrition culture. This has led to a search for healthy foods and local nutrition has begun to gain momentum again (Erdem and Akyürek, 2017: 118).

Warde and Martens (1998) argue that food and beverage services are part of consumers' daily routine, which maintains consumption culture. Today social media as a significant part of consumers' lives is maintaining consumption culture itself with food and beverage influences. With the development of technology, the food and beverage sector has become more visible and more accessible to consumers (Holmberg, 2014). Furthermore, consumers have become more sensible about the influences of social media.

Besides social media being very important for personal communication for many people, digital sharing platforms also reshape the food and beverage culture. This effect derives from sharing by users from different food cultures and cooking styles on digital platforms (Tuç and Özkanlı, 2017).

Consumers are divided into market segments considering that they have similar behaviours in the research, purchasing and experience processes. This distinction is classified according to geodemographic characteristics. Consumers may consider different features when evaluating a food and beverage service. For example, one segment may consider prices, while another segment may prioritise the service's quality or location (Johns and Pine, 2002). This situation may be seen in the effect of social media influencers. Consumers who use social media may be influenced by influencers in the consumption process. Nowadays consumers prefer to check the food and beverage business social media accounts for comments, photographs or videos before they experience. After the consumers share their experiences, they are in the spotlight of these businesses to target the right marketing segment and have become a part of marketing strategy. With this aspect, it is clear that social media has an indispensable place in consumers' food and beverage consuming preferences (Acar, Çizmeci and Turani 2021).

Method

The aim of the book chapter is to investigate the effect of social media influencers on food and beverage culture. A descriptive analysis was conducted to discover this effect. The social media platform, Instagram was used as a source to gather the data. Instagram is one of the leading social media platforms that is used for sharing information, photographs, videos and communication (Hu, Manikonda and Kambhampati, 2014). The most popular users on Instagram in the food and beverage sector category from Turkey have been selected to analyse. The sampling method consists of statistics from 'Boomsocial' (https://www.boomsocial.com/Instagram/UlkeSektor/turkiye/fenomenler) which is a social media measurement and analysis website. From the website, the first 500 influencers from Turkey are listed by the number of their followers. After that, food and beverage accounts have been searched from this list. The last 50 food and beverage-related posts of 31 influencers were systematically reviewed on Instagram. The data in the study were encoded by two researchers to ensure consistency (Patton, 2002).

Findings

The first most popular 500 influencers on Instagram from Turkey are listed. From this list, influencers related to the food and beverage sector have been investigated (Figure 13.1).

First 500 Influencers

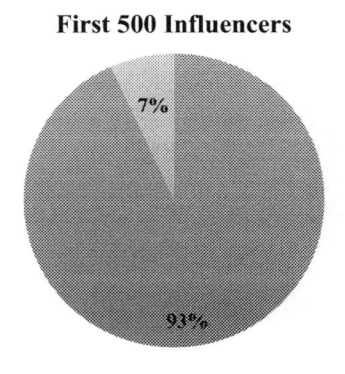

7%

93%

⬛ Total   ⬛ Food&Beverage Influencers

FIGURE 13.1 Percentage of food and beverage influencers in the first 500 most popular influencers.

Among 37 food and beverage influencers, the research sample consisted of 31 influencers because six of them were concept or promotion accounts (Figure 13.2).

Thirteen influencers appeared as 'professional' accounts since they were chefs, restaurant owners or TV show presenters who may assumed to have training (Table 13.1). It has been revealed that most of Turkey's food and

Food&Beverage Influencers

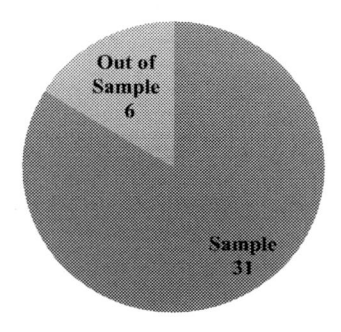

■ Sample ■ Out of Sample

FIGURE 13.2 Sampling of the study.

TABLE 13.1 Food and beverage influencers

	Influencer	Followers		Influencer	Followers
1	Şükran Kaymak (pro.)	6.866.073	16	Özlem Öztürk	1.652.074
2	Ercan Steakhouse (pro.)	4.922.366	17	Refika'nın Mutfağı (pro.)	1.584.434
3	Nermin Yazılıtaş	4.158.984	18	Fulya Çelik (pro.)	1.421.367
4	Nermin Öztürk (pro.)	3.802.138	19	Beyhan Kadayıfçı (pro.)	1.371.540
5	Behiye Kaya	2.781.355	20	Somer Sivrioglu (pro.)	1.363.137
6	Merve Ünal	2.596.988	21	Yasemin Arslan	1.286.724
7	Cahide Sultan	2.467.248	22	Taha Duymaz (pro.)	1.270.230
8	Ayşegül Usluer	2.453.374	23	Gurbet Aygün	1.245.565
9	Zeliha Küçükturan	2.344.587	24	Nuray Akşit	1.179.713
10	Şef Nefel Bulut Yemek Okulu (pro.)	2.201.753	25	Sevim Bekgöz (pro.)	1.099.965
11	Arda Turkmen (pro.)	2.156.064	26	Tahsin Küçük (pro.)	1.038.828
12	Zübeyde Mutfakta	2.111.105	27	Harika Tarifler	1.030.733
13	Fatmanur Keleş	1.920.744	28	Filiz Er	990.989
14	Arif Erdoğan (pro.)	1.747.240	29	Ayşegül Atılgan	963.584
15	Yemek Aşkım	1.730.742	30	Damla Dasdemir	961.278
			31	Nevsin Avdan	869.682

beverage influencers are non-professional individuals sharing homemade recipes from their kitchens.

The distribution of the number of influencers by year is shown in Figure 13.3. A large part of the sample has been using Instagram at least for ten years. The sample consists of 86% women. Professional accounts seem to be distributed equally in terms of gender.

The last 50 posts related to food and beverage of 31 influencers were investigated (Figure 13.4). The posts are classified into 'traditional', 'rival' and 'modern' categories. Analysis showed that 17 of 31 influencers shared mostly 'traditional' recipes referencing classical Turkish cuisine, while 12 of 31 influencers shared both traditional and modern recipes evenly which can be called the 'rival' category. Only two influencers shared mostly 'modern' recipes in their last 50 posts.

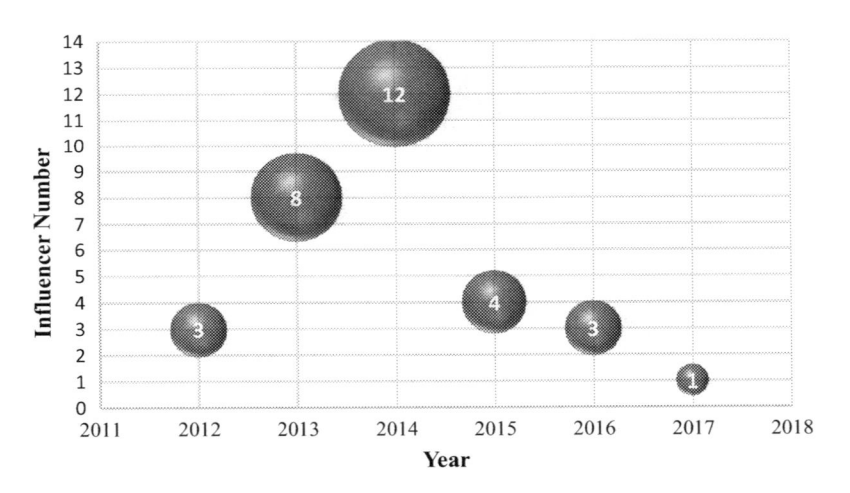

FIGURE 13.3 Number of influencers by year.

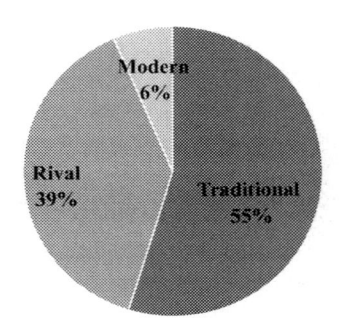

FIGURE 13.4 Last 50 posts of food and beverage influencers categorised.

One of the remarkable findings is the 'rival' category posts scattered as traditional main courses and modern desserts/beverages. Besides, two influencers' posts in the 'modern' category contain mostly desserts, bakeries and beverages like milkshakes, iced tea, etc.

Conclusion

The influence of social media influencers on consumers' food and beverage (F&B) consumption behaviour is a complex and multifaceted phenomenon that has received considerable attention in recent years. In this chapter, key insights and findings were synthesised, recognising the nuanced relationship between social media influencers and F&B consumption behaviour, especially from the traditional cuisine side.

Social media has changed the way consumers interact with, perceive and make decisions about F&B products. Influencers, with their ability to reach large and engaged audiences, have become key players in shaping consumer preferences and choices. One of the most important aspects of influencer impact is authenticity. Consumers often value authenticity in influencer content, which can build trust and credibility. However, the line between genuine recommendations and sponsored content can be blurred, which can undermine trust in influencers. In order to ensure trust, this study was conducted among the first 500 influencers from Turkey. Among these 500, 31 of them were included in the study as food and beverage influencers.

It is important to note that not all influencers have the same impact. Micro-influencers often have more engaged and niche audiences, making them more influential within specific niches. Mega-influencers, on the other hand, can reach a broader audience, but may be perceived as less authentic. For that reason, the sorting of influencers was done according to their number of followers.

The visual nature of F&B products makes them particularly well-suited to social media. High-quality images, videos and aesthetically pleasing content often grab consumers' attention and have a significant impact on their consumption decisions. Moreover, social media has the power to rapidly drive F&B trends. Viral challenges, hashtags and popular influencer recommendations can lead to sudden surges in demand for certain products or dining experiences. According to this study, they have a sort of impact on traditional cuisine through their recipes and posts. It is important to recognise that while social media influencers can generate short-term buzz and demand, their long-term impact on consumers' F&B consumption behaviour can vary. Maintaining consumer loyalty and changing long-term consumption habits is a more complex challenge; therefore, it is important to have a deeper understanding of to what extent they have an impact on traditional cuisine and consumer behaviour on it.

In conclusion, the impact of social media influencers on consumer F&B consumption behaviour is undeniable. It has changed the way consumers discover, interact with and choose food and beverages. However, this relationship is multifaceted, involving trust, authenticity and the power of social media platforms. For societies, food and beverage culture has an important place (Acar, Çizmeci and Turan, 2021). In the field of consumer behaviour, the impact of influencers on traditional cuisine and the ways in which customers are affected by these trends hold significant importance.

Further research is needed to fully understand this dynamic, especially in a rapidly evolving digital landscape. Brands and scholars themselves need to work together to gain a deeper understanding and ensure that influencer-driven F&B consumption behaviour is in the best interests of consumers in traditional cuisine. The complex interplay between influencers, consumers and the F&B industry will undoubtedly continue to evolve, making this an area of ongoing research and study for years to come.

References

Acar, N., Çizmeci, B. & Turan, A. (2021). A research on consumer perceptions of food and beverage marketing on social media. *OPUS International Journal of Society Researches*, *17*(34), 813–830. https://doi.org/10.26466/opus.753599

Aksoy, M., & Üner, E. H. (2016). Rafine mutfağın doğuşu ve rafine mutfağı şekillendiren yenilikçi mutfak akımlarının yiyecek içecek işletmelerine etkileri. *Gazi Üniversitesi Sosyal Bilimler Dergisi*, *3*(6), 1–17. https://dergipark.org.tr/en/pub/gusbd/issue/26626/298960

Brown, D. & Fiorella, S. (2013). *Influence marketing: How to create, manage, and measure brand influencers in social media marketing.* Que Publishing.

Brown, D. & Hayes, N. (2008). *Influencer marketing: Who really influences your customers?*. Elsevier Ltd.

Erdem, B. & Akyürek, S. (2017). Yeni bir mutfak akımı: Yaşayan mutfaklar. *Journal of Tourism and Gastronomy Studies*, *5*(2), 103–126.

Geyser, W. (2023, March 24). What is an influencer?-Social media influencers defined. *Influencer Marketing Hub.* https://influencermarketinghub.com/what-is-an-influencer/

Holmberg, C. (2014, May 5). Food and social media: A complicated relationship. *Huffpost.* https://www.huffingtonpost.com/christopher-holmberg/food-and-social-media-a-_b_4898784.html

Hu, Y., Manikonda, L., & Kambhampati, S. (2014). What we Instagram: A first analysis of Instagram photo content and user types. *Proceedings of the International AAAI Conference on Web and Social Media, USA*, *8*(1), 595–598.

Johns, N., & Pine, R. (2002). Consumer behaviour in the food service industry: A review. *International Journal of Hospitality Management*, *21*(2), 119–134. https://doi.org/10.1016/S0278-4319(02)00008-7

Kadekova, Z. & Holiencinova, M. (2018). Influencer marketing as a modern phenomenon creating a new frontier of virtual opportunities. *Communication Today*, *9*(2), 90–105. https://www.communicationtoday.sk/download/22018/06.-KADEKOVA-HOLIENCINOVA-%25E2%2580%2593-CT-2-2018.pdf

Olenski, S. (2017, September 25). The impact of live streaming on influencer marketing. *Forbes.*   https://www.forbes.com/sites/steveolenski/2017/09/25/the-impact-of-live-streaming-on-influencer-marketing/?sh=2c9c34afe607

Patton. M. Q. (2002). *Qualitative research and evaluation methods (3rd ed.).* Sage Publications.

Riley, M. (1994). Marketing eating out: the influence of social culture and innovation. *British Food Journal, 96*(10), 15–18. https://doi.org/10.1108/00070709410072463

Tuç, Z., & Özkanlı, O. (2017). Resource about reshaping of food and beverage culture by social media: Sample Gaziantep City. *Journal of Urban Culture and Management, 10*(2), 216–239.

Türker, N., & Süzer, Ö. (2022). Tourists' food and beverage consumption trends in the context of culinary movements: The case of Safranbolu. *International Journal of Gastronomy and Food Science, 27*, 100463. https://doi.org/10.1016/j.ijgfs.2021.100463

Warde, A., & Martens, L. (1998). Eating out and the commercialisation of mental life. *British Food Journal, 100*(3), 147–153. https://doi.org/10.1108/00070709810207513

Ye, G., Hudders, L., De Jans, S. & De Veirman, M. (2021). The value of influencer marketing for business: A bibliometric analysis and managerial implications. *Journal of Advertising, 50*(2), 160–178. https://doi.org/10.1080/00913367.2020.1857888

14

INDUSTRY VIEW

Reimagining customer experience in the hospitality and tourism industries: Navigating evolving values and expanding authenticity in the post-pandemic landscape

Michael Donald

Introduction

The allure of travel has always drawn human interest, tempting us to explore new places, immerse ourselves in different cultures and expand our horizons. The sense of adventure feeds our collective souls. From highway inns to underground hipster cocktail bars, the hospitality and tourism industry is a silent partner in this quest and a facilitator of innumerable special memories. At the core of this dynamic industry have been the principles of the customer experience and authenticity – not just as business strategies but as the creators of bonds that emotionally connect travellers to places, people and brands.

The COVID-19 pandemic brought unforeseen challenges, shaking the foundations of movement and travel upon which the hospitality and tourism industries were built. It necessitated a hurried recalibration of norms and practices as borders abruptly closed and health protocols came to the fore, prompting a dramatic shift in customer priorities and expectations that have continued to evolve since then.

Guest loyalty has always been a crucial factor in this industry, fostering sustained business revenues and long-term relationships with customers. These relationships have historically been the key to reducing operational costs while simultaneously enhancing revenue streams, nurturing a cycle of trust and preference that goes beyond mere transactional interactions. However, loyalty could often be trumped with an enticing discount, and the resulting price wars always lead to a squeeze on margins. Today's discerning customers are not just seeking services but aspiring to build strong and enduring relationships with brands that resonate with their values and preferences. These connections are

DOI: 10.4324/9781032637778-15

likely to withstand the test of time and present opportunities for the hospitality industry to build on long-standing loyalty that is less distracted by price.

This chapter seeks to forge a path through this redefined landscape, dissecting the shift in customer expectations and values from a pre- to a post-pandemic world. It aims to spotlight the enhanced focus on guest loyalty and authenticity, not just as a business strategy but as an essential pillar in building strong, long-lasting relationships with customers who now yearn for more authentic engagements and deeper connections with the brands they patronise. For the purposes of this chapter, the author has interviewed four hospitality leaders from various sectors of the industry: David Gardner is a Managing Partner at 80 Days, a creative digital marketing agency that specialises in hospitality and tourism: Andrew Kennedy is the Regional Director Franchise Owner Support for Marriott International, one of the world's leading hotel companies; Rob Flinter is the General Manager at Park Plaza Waterloo and Andrea Shaw is a Director at FM Recruitment. Each interviewee was presented with a questionnaire to explore their experiences regarding customer experience, guest expectations, technological advancement, recruitment and organisational values. The author used a thematic method to identify trends and presented these findings to support the article. All quotes from these interviews will be referenced (Donald, 2023).

Historical context and consumption trends: Setting the pre-pandemic stage

Prior to the unprecedented upheavals brought about by the global pandemic, the hospitality and tourism sectors were ingrained in a landscape largely influenced by the perceived value of budget-friendly options. Serviced apartments, Airbnb, select-service hotels and hostels have all enjoyed healthy growth over the last decade. In this setting, a competitive market was crafted to serve a consumer demographic predominantly swayed by economic factors, constantly in search of the most valuable experiences moulded by the global currents of economic and travel narratives.

Although elements of personalisation and authenticity were not absent during this period, they were largely overshadowed and relegated to the background by the overpowering emphasis on price. As a result, customer experiences were mainly dictated by tangible factors such as the convenience of the location, the availability of amenities, encouraging reviews and, notably, the financial viability of choices. The industry landscape was largely transactional, where a strong focus on cost sometimes overrode the potential for deeper, more enriching engagements.

However, this period also saw the early stages of data-driven, technology-enhanced guest loyalty strategies, with businesses starting to grasp the mutual benefits of fostering enduring relationships with their customers. Loyalty began to represent a pathway to optimised revenue streams, encouraging recurring business while seeding a culture where customers yearned for a deeper connection with their favourite brands. This laid the foundation for a shift towards a relationship-centric industry landscape.

As we delve further into this historical background, we scrutinise a timeframe where the mindset of travellers was considerably angled towards the pursuit of value, frequently placing a greater emphasis on cost-effectiveness over the richness of authentic experiences. Understanding this baseline is pivotal in appreciating the substantial shifts currently unfolding in the hospitality industry in the post-pandemic period – a movement leaning towards a balance that honours guest loyalty while envisioning a landscape where both businesses and customers are eager to nurture relationships rooted in authenticity and enriched personalisation. This current trend sets the stage for fostering genuine connections and crafting experiences that resonate on a deeper, more meaningful level, ushering in a promising future for the sector with a wealth of opportunities yet to be explored.

The evolution of consumer values in a post-pandemic world

In the aftermath of the pandemic, a marked transformation in consumer values within the hospitality and tourism sectors has materialised, significantly altering industry trajectories. Far beyond health implications, the pandemic has instigated an overhaul in consumer values, touching on aspects such as time allocation, effort and social interaction. This change has sparked a drift towards value-seeking behaviour, towards a discerning preference for experiential value and authenticity. In her article 'Opportunities for tourism due to post-pandemic travel triggers', Dr Cindy Heo comments that 'the pandemic did not fundamentally change the major attributes that affect tourism destination choice (i.e., natural scenery and culture) and an ultimate value of travel (i.e., life enrichment). *Learning* was addressed as a key functional benefit of travel, whereas *experience* was stated as the most important emotional benefit of travel'[1] (Heo 2023).

Travellers, now more informed and cautious, are willing to invest premiums for experiences that forge deeper connections with the places they visit and the cultures they encounter. We have become more risk-averse and unwilling to gamble our time and money on experiences that might not fulfil us. Broadly speaking, the hospitality industry comprises of four main sectors: Accommodation; Food & Beverage; Travel & Tourism; and Entertainment & Leisure.[2] Let us assume that customers in each of these sectors would share similar values. In the theatre world, we can see how new stage productions being dominated by productions inspired by cinema. Likewise, the film industry has become dominated by sequels, spinoffs and comic-book-inspired cinematic universes. We often see individual films and TV series playing an episodic role in a much larger piece of work that also requires a more significant time commitment from the viewer to complete. Consumers feel safe within the familiarity of stories they recognise, and in accommodation, food and beverage we see similar trends as globally franchised operations offer familiar and consistent products and services that continue to drive loyalty. This shift offers a golden opportunity for brands to cultivate stronger guest loyalty by fostering relationships built on mutual trust and a shared appreciation for quality, thereby creating brand advocates and driving revenue while minimising acquisition costs.

According to a *Harvard Business Review* article[3] (Gallo, 2014), increasing customer retention rates by 5% can increase profits by between 25% and 95%. In a post-pandemic era with the cost-of-living crisis and high inflation, the question of building loyalty is not just a mark of pride but has a direct and substantial impact on the bottom line. Return customers are well known to help reduce marketing budgets and the costs involved in winning new customers. There is even more significant profit protection should guests book directly, but loyal customers can also spend more, especially when they stay longer and more regularly.

Andrew Kennedy, Regional Director Franchise Owner Support for Marriott International, commented,

> Marriott Bonvoy (Marriott's loyalty programme with just over 190 million members globally) guests are not only more loyal, they often pay premium rates in order to access the higher tier benefits and points. While the pandemic had a far more severe impact on the business, we saw similarities with the 2008 financial crash. Namely, our loyalty programme (Starwood Preferred Guest at the time) was essential to keeping a larger share of the customers that were travelling, but it was even more crucial in helping us to rebound once guests were able to travel again.[4]
>
> *(Donald, 2023)*

The evolving landscape reveals a reimagined tourism model, characterised by a clear divergence from mass tourism towards more personalised and adventurous experiences that echo the location and sense of place. The rise of boutique hotels, niche tour groups, and a baying audience on social media epitomises this trend, revealing hidden gems to those in pursuit of authenticity. With 31 brands and more than 8,500 properties across 138 countries, few hospitality groups are better placed than Marriott to identify international trends as they evolve. Andrew adds:

> While not all of our brands are franchiseable, Tribute Portfolio and Autograph Collection have been gaining popularity amongst guests, operators, and owners because they offer unique experiences while still being

connected to the larger Marriott brand. While a Sheraton or Marriott has set brand standards where walls must be painted a certain colour and particular scents must be in the lobby, Tribute Portfolio and Autograph Collection are designed to identify the core branding elements for individual properties. This gives owners and operators the flexibility to create unique properties that reflect the local culture and values, which appeal to that new generation of guests looking for authentic and/or Instagrammable experiences.[5]

(Donald, 2023)

Within this revamped narrative, guest loyalty emerges as a fundamental pillar, symbolising a harmonious relationship between consumer and brand. There has certainly been a trend for an increased emphasis on guest loyalty in the overall guest experience over the last decade[6] (Kim, Vogt & Knutson, 2013). This dynamic paves the way for collaborations that honour long-term shared values and mutual respect rather than individual moments of satisfaction or short-lived delight, thereby fuelling business prosperity and enriching customer experiences.

As businesses recalibrate to align with evolving consumer sensibilities, the focus is honed on crafting experiences that resonate personally, nurturing enduring relationships and fostering a landscape ripe for discovery, enrichment and genuine connection. This change signals a meticulous approach to choices, influenced by factors like cleanliness standards, crowd sizes and booking flexibility. The resultant surge in demand for destinations and establishments that can offer reassurance on these fronts marks an era where quality and safety are prioritised.

Post-pandemic travel has transcended the realms of escape and leisure to intertwine with a quest for meaning, rejuvenation and assurance. Modern travellers are drawn to destinations offering spacious, natural environments and experiences that resonate on a deeper emotional level through wellness retreats, nature immersion or cultural engagements. The dining scene has similarly evolved, with a marked preference for local produce, sustainable practices, and health-centric menus and a penchant for airy, open spaces or alfresco dining, over confined interiors.

Socialising patterns have been reshaped, with a renewed appreciation for genuine connections, resulting in a preference for intimate gatherings and personalised events that foster meaningful engagement. There's an augmented emphasis on authenticity, transparency and ethical considerations. Consumers are keen to understand the ethos behind the services they engage with, demanding forthrightness regarding sourcing practices, sustainability measures and assurances that employees are well taken care of.

Establishments that effectively communicate their ties to the local community showcase farm-to-table practices, prioritise genuine eco-tourism, proudly and publicly pay a 'living wage', and/or have transparency on how gratuities are distributed amongst the team resonate more profoundly with consumers today. These genuine narratives have become significant differentiators in a market where customers seek more than superficial value, signalling a fundamental shift in the consumer value landscape of the hospitality and tourism sectors in a post-pandemic world.

The role of technology in this new landscape

Amidst the upheaval of the pandemic, technology firmly positioned itself as an indispensable ally for the hospitality industry. Technological advancements not only took significant strides but proved pivotal in enabling businesses to navigate the particular challenges they faced. From streamlining operations to offering seamless virtual experiences, technology was instrumental in ensuring continuity and resilience. As we settle into the post-pandemic world, tech solutions are becoming far more tailored to the real-life challenges and opportunities in the industry. This responsiveness to hospitality needs is evident in the emergence of more flexible contract models and reduced upfront capital expenditures. Additionally, with the ever-increasing prevalence of Open APIs (Application Programme Interfaces) that allow different technologies to integrate seamlessly, technology partners can take the manual labour out of eliciting feedback from customers. This helps hospitality businesses engage with customers to understand them better, and review and adapt their processes to readily align with rapidly evolving market conditions. 'Feedback either positive or negative, by guests signals a willingness to engage with the brand and increases the probability of them becoming loyal by almost 50%'[7] (Anderson & Saram Han, 2021).

Rob Flinter, General Manager at Park Plaza Waterloo, encapsulates this sentiment aptly, highlighting the significance of agility and foresight in technological collaborations. He states,

> In terms of choosing the right technology partners, the crucial issue for us (Park Plaza Hotels & Resorts) is to stay flexible and be ready to adapt to changing situations should demand or demographics change. Getting locked into long contracts, huge capital expenditures, or systems that can't evolve with us and our customers are considerable red flags.[8]
>
> *(Donald, 2023)*

This perspective underscores the importance of technology not just as a tool but as a dynamic partner, consistently gauging and adjusting to shifts in customer values and the broader hospitality terrain.

However, for these technologies to really succeed, it relies on the hospitality industry truly understanding their customers and ensuring that every touchpoint in their journey adds value to the experience. Revinate, a market-leading Customer Relations Management (CRM) platform, suggests that there are 26 touchpoints on the customer journey, and each one of these has a cumulative effect on the customer experience, directly impacting how potential customers feel, the decisions they make and the long-term loyalty that may result. While first-person marketing data has obvious advantages for a business, if the connections made don't feel like authentic efforts to improve the guest experience, the modern customer will be far more inclined to unfollow, unsubscribe or simply not book again. At the same time, if technology is utilised to enhance the customer experience, anticipating needs and giving control and choices to customers, it can act as a powerful tool that can be utilised, refined and improved upon to find better and more innovative solutions for customers and businesses alike. Rob Flinter adds:

> We use Revinate to collect and collate guest feedback. We recognise that there is not just a great opportunity for the business but a responsibility to our customers to ask the right questions and then act upon the feedback they share. We have regular guest-focused meetings in the hotel to take feedback from the guests and work as a team to find opportunities to continually improve our product and service offering to ensure that our customers choose to come back to stay with us or another hotel in the brand time and time again.[9]
>
> *(Donald, 2023)*

The future

With the cost-of-living crisis, inflation and Brexit all contributing to inevitable price increases, the modern consumer, equipped with a revamped set of priorities, seeks more than just monetary savings. On top of these, hospitality businesses in the UK have faced a recruitment and retention crisis that has taken away the option of dropping prices to increase volume. While this tactic proved popular during the financial crisis in 2008, hospitality was relieved not to enter a post-pandemic price war this time, which would have stripped profit and possibly sacrificed long-term brand loyalty and value creation. As the focus on price recedes, four pillars emerge as critical cornerstones:

Quality: No longer content with sub-par offerings, consumers demand top-notch services, impeccable hygiene standards, and facilities that resonate with luxury and comfort, regardless of the price bracket. Instead of sheer volume, many successful businesses are putting more focus on guest satisfaction, repeat visits and positive word-of-mouth. This may entail refining service offerings, investing in staff training, or even rethinking the design and ambiance of spaces to align with modern sensibilities.

Authenticity: Today's traveller is an explorer at heart, seeking genuine experiences. Be it a locally sourced meal, a culturally immersive workshop or an authentic local stay, the quest is for something real, unfiltered and memorable. Post-pandemic trends such as the rise of the digital nomad (where employees or freelancers can work from anywhere in the world), as well as the 'bleisure' trend (where leisure trips are being peppered with working days), are allowing travellers to spend much more time in far-flung destinations, becoming far more in tune with the local culture and lifestyle and consequently experiencing a much more authentic perspective. Businesses can learn from this and prioritise genuine, personalised engagements over generic offerings. Be it through tailored travel packages, immersive cultural experiences or simply recognising and appreciating customer preferences, making each guest feel seen, heard and valued will always help connect a brand with their customers.

Sustainability: With a growing global consciousness about environmental issues, there's a surge in demand for ecofriendly travel options and a growing intolerance for greenwashing. Consumers are keen to see a credible and holistic strategy for sustainable hospitality rather than a couple of actions paying lip service. Paper straws in the bar will no longer be able to paper over the use of plastic single-use amenities in a hotel room. The future of sustainability in hospitality will be at the heart of operations. Whether it's sourcing locally, minimising waste or investing in green technologies, businesses that actively and transparently showcase their commitment to the planet will resonate with environmentally conscious travellers and also reap long-term cost savings and operational efficiencies.

Experiential Value: Beyond mere accommodation or travel, consumers are looking for transformative experiences – journeys that touch the soul, challenge perspectives or offer unique stories and connections. Our internal customers are no different. A fulfilled and well-cared-for workforce is the cornerstone of outstanding guest experiences. Organisations should prioritise employee well-being, be it through competitive remuneration, growth opportunities or mental health initiatives, but most importantly by fostering a culture of respect and inclusivity in the workplace. Andrea Shaw, Director at FM Recruitment, states that 'toxic workplace culture is now the number one reason employees give for moving jobs. This is followed by low salaries, but we often see employers focusing more on the salary issue than addressing opportunities for improving workplace culture'[10] (Donald, 2023).

Conclusion

As we conclude our assessment of the evolving dynamics of the hospitality industry, distinct themes emerge. The post-pandemic era isn't just a revival of the old, but an evolution, marked by shifting values, an amplified yearning for authenticity, and a deeper pursuit of meaningful interactions.

The hospitality industry's proven resilience, demonstrated by its rapid rebound from the unprecedented challenges of the COVID-19 pandemic, is a testament to its inherent capacity for adaptability, growth and innovation. This resurgence has been powered by a pent-up demand and the timeless human need for connection and discovery.

The current trend doesn't necessarily indicate a complete departure from price-focused decisions. Customers will always factor in price before committing to a purchase. Instead, there's a significant opportunity for the hospitality sector to shift away from relying heavily on price cuts to draw customers. A strategy that more effectively conveys core values, creates experiences that really resonate and forges enduring connections could help tip the balance. Guests in this new era are not just consumers; they are in search of moments that touch their souls, tales that remain, and bonds that last.

Technology emerges in this new chapter not just as a facilitator but as a strategically, creating richer, more personal connections between establishments and their guests. However, even as technology brings about remarkable transformations, the essence of true hospitality – the sincere desire to serve and connect – should remain unaltered. The future of this dynamic industry, guided by authenticity and powered by adaptability, will be crafted by a harmonious blend of innovation and the timeless warmth of genuine hospitality.

Notes

1 Heo (2023). https://hospitalityinsights.ehl.edu/opportunities-for-tourism-and-hospitality.
2 Birmingham Business School. https://www.bcu.ac.uk/business-school/blog/the-different-hospitality-sectors-explained.
3 Gallo The Value of Keeping the Right Customers https://hbr.org/2014/10/the-value-of-keeping-the-right-customers.
4 Donald (2023). Interviews conducted between the author and hospitality leaders to support this article.
5 Donald (2023). Interviews conducted between the author and hospitality leaders to support this article.
6 Kim, M., Vogt, C.A., Knutson, B.J. Relationships among customer satisfaction, delight, and loyalty in the hospitality industry. *J Hosp Tourism Res*. 2015;39(2):170–97. 9. https://www.researchgate.net/publication/276077892_.
7 Anderson & Han, Customer Engagement: The Key to Long-term Loyalty and Impact, Cornel Centre for Hospitality Research, https://ecommons.cornell.edu/server/api/core/bitstreams/5e2f16c7-baf4-4b06-90aa-7c9eb4140d26/content.
8 Donald (2023). Interviews conducted between the author and hospitality leaders to support this article.
9 Donald (2023). Interviews conducted between the author and hospitality leaders to support this article.
10 Donald (2023). Interviews conducted between the author and hospitality leaders to support this article.

INDEX

Printed in the United States
by Baker & Taylor Publisher Services